The Lone Wolverine

The Lone Wolverine

◆ ◆ ◆

*Tracking Michigan's Most
Elusive Animal*

Elizabeth Philips Shaw
and
Jeff Ford

The University of Michigan Press ◆ *Ann Arbor*

Published in the United States of America by
The University of Michigan Press
Manufactured in the United States of America
⊛ Printed on acid-free paper

2015   2014   2013   2012   4   3   2   1

A CIP catalog record for this book is available from the British Library.

ISBN 978-0-472-11839-7 (cloth : alk. paper)
ISBN 978-0-472-03487-1 (pbk. : alk. paper)
ISBN 978-0-472-02847-4 (e-book)

To my children, Summer, Courtney, and Nathan, for all the years of love and support; to their loved ones, who have grown our family in such wonderful ways; and to my parents, who always believed. Thank you.

—ELIZABETH PHILIPS SHAW

✦

To my father, Jac, and Grandpa Charlie for instilling in me a passion for the outdoors; my mom, JoAnn, for taking me in as one of her own; my sister, Teri Skidmore, for designing the website and maps for this book; my buddies, Steve Noble and Jason Rosser, for their help with research; Carol Dusek, Marilyn Millikin, Betty Trumble, Laurinda Rose, and Barbi Kennedy for their love and support; Aaron and Ryan Schenk for not allowing the wolverine to be shot; Steve McNiel for use of the spy cam and video system; my Deckerville family, which made teaching very gratifying; my father-in-law, Harold Titus, for nearly two decades of hunting bucks together with stick and string; Audrey Magoun and Judy Long for keeping the *Gulo* fire burning deep inside me; my wife, Amy, and children, Riley and Clint, for their love and support; and Elizabeth Shaw for her love and interest in the wolverine, her writing skills, and for being such a positive force in getting this unique story into print!

—JEFFREY JAC FORD

Preface

Elizabeth Philips Shaw

✦ ✦ ✦

Back in the spring of 2010, I had recently left a 12-year post as a reporter for the *Flint Journal* in Flint, Michigan, where I worked as the outdoors, environment, and health beat writer. I had, of course, read occasional news stories over the years about Michigan's Thumb wolverine, but my first real encounter with the story wasn't until shortly after the wolverine's death in March 2010. A friend and fellow outdoors writer, Tim Lintz, called to tell me the news and suggested I might like to contact Jeff Ford and others involved.

I knew it would be a good story. What I didn't expect was to be so moved by this gruff-spoken high school science teacher and football coach—and by the obvious love and devotion he felt for this wild animal he had studied from afar for six years, for no reason or reward beyond his own passion and innate curiosity.

By the time our first phone conversation ended, I knew this was a book I wanted to write, and I am honored and grateful that he has allowed me to write it with him.

Although we have taken great pains to ensure its scientific accuracy and have cited all academic references, this is not a scientific or academic treatise on wolverines. There are far better books of that nature—by some of the very researchers cited here, in fact.

This is simply the true story of one mysterious, misplaced wolverine—Michigan's own last unicorn—and the man who devoted his life to her. It is a love story as odd and rare as the wolverine herself and just as unlikely to ever pass this way again.

Acknowledgments

✦ ✦ ✦

The authors would like to thank Dr. Audrey J. Magoun of Wildlife Research and Management in Fairbanks, Alaska; Judy Long, Administrative Manager of The Wolverine Foundation, Inc.; US Forest Service Wildlife Biologist (retired) Jeffrey P. Copeland; and Michigan Department of Natural Resources Wildlife Biologist (retired) Arnold Karr for their invaluable input and assistance in assuring the scientific and factual accuracy of this book. We would also like to thank John Schmittroth for his wisdom and guidance in the project's earliest stages.

Contents

✦ ✦ ✦

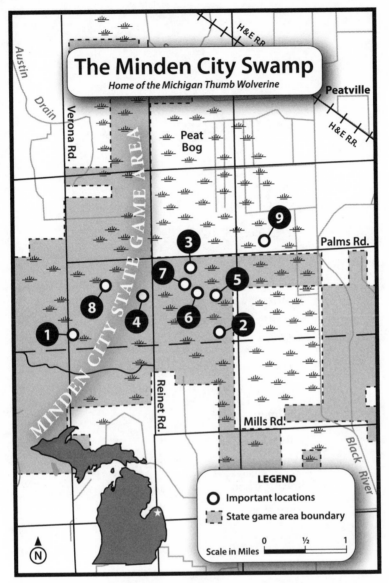

Map 1. Minden City State Game Area

1. Abbott comes across tracks (winter 2003)
2. Ford and Noble find their initial tracks, March 6, 2004
3. Ford finds the wolverine's tracks in the mud on June 6, 2004, and makes casts, which allows the men to confirm that this is indeed a wolverine
4. Ford tracks the wolverine along the eastern end of the canals on March 12, 2004
5. Ford takes the first picture of the wolverine on March 12, 2005
6. The wolverine charges Ford and Rosser on April 10, 2005
7. Research site where Ford takes the majority of the pictures and video over the five-year period
8. The canals, the wolverine's sanctuary
9. Rann and Graham find the wolverine dead on March 13, 2010

March 13, 2010

✦ ✦ ✦

It was still a week out from spring and unseasonably warm for a state as far north as Michigan, especially in the mitten-shaped state's Thumb region—an endless place of flat farm fields and narrow woodlots, where every scattered, solitary farmhouse and silo rears up from the land like the lonely hand of a drowning man thrust from a waveless sea. The roadside ditches are cut dead straight and deep enough to carry a flood or lose a truck, stretching off to the horizon like raw, surgical gouges in the earth. This is a place where wind farms are built . . . where coyotes and combines run. A place where nothing stops the wind.

On this particular mid-March day, a warm east wind was blowing across the Thumb, gusting up to 25 mph as it chased off the last icy dregs of winter with a steady drizzle of rain.

All in all, it was a perfect day to do some scouting for the next fall's deer season, thought Todd Rann, an avid bowman who often hunted at the Minden City State Game Area, square in the region's midsection, halfway between Port Huron and Bad Axe, about a half hour's drive from the Lake Huron shore.

The remote, swampy state land is thick with bugs and underbrush and attracts few casual hikers. Rann and girlfriend Morgan Graham had the entire place to themselves that early pre-spring day. They were hiking along the main trail through the Minden bog when the cell phone in Rann's backpack rang.

While he paused to answer it, Graham wandered off the trail a few dozen yards to check out a nearby beaver dam. Rann himself had passed the dam dozens of times before, so it held little interest for him anymore. Most days he never even bothered to glance in its direction when he hiked past on his way to somewhere else.

He was just finishing his phone call when he heard Graham call out that she'd found something dead in the pond.

Curious, he walked over to join her at the water's edge, about 30 feet from the beaver dam.

"I looked down and saw this big, brown, dead animal floating in the water. I couldn't get hold of it, so I grabbed a log and pushed it close to the bank, then leaned down to grab it out," said Rann. "I was pretty surprised when I pulled it out."

In fact, it took several seconds for his mind to register what he might be holding: the wet, half-frozen carcass of what appeared to be a wolverine.

Rann had seen all the newscasts over the past few years, read the newspaper stories. But was it possible that this was indeed Michigan's famous lone wolverine?

Despite Michigan's self-proclaimed title as the "Wolverine State," most authorities had considered the wolverine (scientific name *Gulo gulo*) absent from Michigan's ecology for at least 200 years, ostensibly wiped out by trappers and habitat loss.

The cunning, ferocious predator isn't related to the wolf at all and is instead a member of the weasel family. In appearance, it looks roughly like a cross between a badger and bear, with a temperament more fitting to an animal several times its size. Typically growing to a mere 25 to 35 pounds, wolverines are reclusive creatures that can erupt into a clawed ball of fury when threatened and have been known to fight off challengers as large as a wolf or bear.

Once widespread across the northernmost tier of the United States from Maine to Washington State, the wolverine's range is now mostly limited to the boreal forests of northern Canada and Alaska, with small self-sustaining populations in Idaho and Montana.[1]

Prior to 2004, Michigan's last known sighting was by fur traders around the turn of the eighteenth century, years before the territory officially became a state in 1837. [AUTHOR'S NOTE: That long-standing bit of historical trivia might need to be revised. In December of 2011, a *Newberry News* staffer unearthed this December 29, 1911, *Newberry News* brief for the Upper Peninsula paper's "Traveling Through Time" column: "Henry Brandt, a trapper, captured a wolverine in one of his traps last week. The animal had dragged the trap several miles in an effort to escape. This is the first instance on record of one of these animals ever being captured in this section."[2]]

Of course, the physical lack of wolverines has never stopped Michiganders from embracing the animal as their unofficial symbol. "Wolverine State" might not be the motto branded on license plates, but you'd be hard-pressed to find anyone in Michigan who doesn't recognize it as part of his or her identity.

The truth is, no one is entirely certain how the whole thing started in the first place. The most popular versions claim the nickname was invented by feuding Ohioans to describe their northern neighbors as particularly vicious and ornery.[3]

Indeed, University of Michigan historians proudly point to the 1835 Toledo War as the rightful roots of the university mascot.

According to the university's own website, "Bad blood from this incident has persisted to this day. The Ohioans began calling the Michiganians 'Wolverines,' the ugliest, meanest, fiercest creatures from the north. In short, the people of Michigan were not to be trifled with."[4]

The fact that their state's powerful totem has been relegated to the status of ancient legend has long been a source of wistful angst for many Michiganders. Over the years, every unconfirmed sighting has caused a flurry of intense media interest and public speculation bordering on the passion of ufologists.

When a group of coyote hunters spotted an unusual lone animal in a rural area about 90 miles north of Detroit on February 24, 2004, the contact escalated into a full-scale mass hunt. Within hours, a state wildlife biologist was on the scene and had positively identified it as the first and only wild wolverine authenticated in the state in the past two centuries.

The announcement made headlines immediately. The rare and elusive creature became an object of intense curiosity for hunters, naturalists, wildlife biologists, and nature lovers across the state and beyond. Was it male or female? Where had it come from? How did it get here? Was it alone?

The speculation and debate continued to rage over the next six years. Was it a true wild disperser that had wandered in from established populations in Canada—or someone's escaped exotic pet?

Was it even the first wild wolverine to be fully documented in the state since 1911? According to some sources, another wolverine had been shot and killed in 1932 in the rural dead center of the Lower

Peninsula. At the time, state officials concluded it most likely had escaped from a zoo, and the case was quickly forgotten by all but a handful of die-hard believers. Today no physical evidence remains from which to draw a definite conclusion either way.

But finally, in 2004, there was a bona fide specimen alive and well and maintaining itself quite nicely in the wild, the riddles of its origin ripe for the picking. For many, it was more than a wildlife mystery. It was a mythological creature from the distant past, a legend come to life: Michigan's own last unicorn.

And now, six years later, she was dead. Her appearance and continued existence had always seemed a bit like a miracle to many. Now it seemed like another miracle that anyone had found her remains at all in the vast, forested swampland she'd long claimed as her domain.

But for Todd Rann and Morgan Graham, it was a simple twist of fate on a gray and rainy March day.

"I knew of it, but I didn't pay a lot of attention and certainly wasn't looking for it that day. I just remembered there being one in the area a few years back so I figured it was the same one," said Rann. "But what were the chances anyone would even be out there to find it, or that my cell phone would ring right as we were walking by? It's like we were supposed to find it."

Graham called a girlfriend, who looked up the number for the Report All Poaching hotline manned by the Michigan Department of Natural Resources (DNR). They dragged the body to a nearby bush and walked back to the parking lot to wait for the conservation officers to arrive.

A half hour later, they led the team back to the site, where the officers confirmed what the couple already knew in their hearts: Michigan's only known wild wolverine was dead.

Schoolteacher Jeff Ford was home alone with his five-year-old son Clint when the call came from a buddy, Steve Noble, telling him a pair of hikers had found the dead wolverine partially submerged in a swamp in the Minden City State Game Area, about an hour from Ford's Cass City home. It was 5:32 p.m.

Since the day the coyote hunters had first spotted her six years before, the wolverine had been Ford's singular obsession. For the first 370 days he'd tracked her through mud and snow and rain—not willing to give up until he could see for himself that Michigan's wolverine was real. On the 371st day he finally succeeded in getting his first photograph, deep in the bowels of the Minden bog.

That might've been enough for most men. It wasn't nearly enough for Ford.

For the next five years, he followed her trail until he knew her habits as well as his own, eventually working with wildlife biologists and conservationists around the country in his quest to fully document this genuine wildlife treasure. His photographs, videos, and physical samples added substantially to the resources of the wolverine science community and provided a unique and vivid window into the hidden life of this extraordinary creature.

In failing health due to a serious heart condition, Ford had stubbornly risked his life numerous times in pursuit of his wolverine, willfully ignored family and friends, exhausted funds and dug himself into deep debt, all for the sake of his self-appointed life's mission.

And now she was dead. Ford slowly hung up the phone.

Steve had told him he was meeting a conservation officer at the corner of M-19 and Deckerville Road in 15 minutes, so Jeff knew he needed to move fast. He quickly gathered up Clint and jumped in his truck. He was breathing hard, almost hyperventilating, his heart racing wildly.

He stopped short when he saw the frightened look in his young son's eyes.

"He knew something was terribly wrong from the way I was acting," said Ford, recalling the events of that day. "I got my breathing back under control. Then I told him we were going to see a dead wolverine."

Ford and Noble caught up with the Department of Natural Resources team at a rural intersection en route to the wildlife lab at Michigan State University where an autopsy could be performed while the remains were still fresh. Ford was no stranger to them. They stopped long enough to let the two men climb into the back of the truck bed where the body lay.

"I thought of all the video I'd taken of her over the years, all the hours watching her playing, rolling around in the snow, climbing trees, just having fun," Ford recalled. "I never dreamed the day would come when I'd be able to actually touch her."

Jeff Ford holds the dead wolverine an hour after the animal was removed from the Minden City State Game Area by the DNR on March 13, 2010.

He didn't cry. Not yet.

With the efficiency of hunters long practiced at field-dressing game, they quickly searched her body for any obvious signs of an injury that might have led to her death such as bullet holes or trap marks. There was nothing. At Jeff's request, Steve snapped a few pictures of him solemnly cradling the body in his arms like a drowned child.

They talked to the conservation officers a few minutes more, then they parted ways. Jeff drove home in silence, Clint riding next to him. He called his wife, asking her to meet him at home, and quickly dropped off their son. He didn't have to explain. Then he drove deep

into state land, parked his truck, and opened a bottle of Busch beer—
the first of many—as Iron Maiden's 1984 "Remember Tomorrow"
played on the radio.

He sat there, beer after beer, slowly getting drunk, listening to mu-
sic as the tears finally came. It was many hours later when he turned the
key in the ignition and headed back home alone.

Had it really been six years since he'd first begun to track her? He
could still see in his mind those first vague prints he and Steve Noble
had found in the melting snow: the toes indistinct, their placement uncertain—but the first solid clue that the object of his pursuit was close
enough to be real.

It was only the next step forward in a lifelong metamorphosis, from
lonely child to troubled loner, then to devoted teacher and father . . .
and now, six years later, to a man known by thousands as simply the
Wolverine Guy.

"The day the wolverine died and the week or two after were terrible. I thought I might shed a tear when she was gone, but I had no idea
what an emotional wreck I would be," said Ford. "I couldn't understand
at the time and still don't even today, how I became so attached to an
animal I could never touch or pet.

"The vast majority of people wouldn't shed a tear over the wolverine, but then again, the vast majority of people wouldn't hump 1.5 miles
out to a swamp going in snow and muck up to their knees with a 40-
pound pack on their back to research her for five years straight.

"So I guess I'm glad that I can cry over her because without that
there isn't passion. Without the gift of passion, the motivation to en-
dure, to persevere and to overcome obstacles would be lost . . . and it all
would have never happened."

December 18, 1970

✦ ✦ ✦

Jeff Ford was born in 1965 in Saginaw, Michigan, the youngest child of Jac and Barbara Ford. Both were athletes and competitive archers with a reputation among their fellow sportsmen around much of the state.

"That was their hobby. They practiced together and went to competitions together regularly. They both won dozens of trophies, and my dad was runner-up to the state champion in the clout shoot one year," recalled Ford.

But Jeff, the youngest of their two children, was far from strong and athletic.

Born with a condition known as pyloric stenosis, Jeff's stomach was plugged at the opening to the small intestine, making it impossible for the tiny infant to do anything but vomit up any nutrition. He was literally starving to death in his mother's arms. Emergency surgery repaired the problem and Jeff recovered fully, but Barbara couldn't get over it. To her, he would always be the frail, fragile infant she'd nearly lost. She became fiercely overprotective of her young son, unwilling to even let him play outside for fear he might get sick or injured.

All that changed one ordinary winter day.

The date was December 18, 1970. Jeff was five years old. Mother and son were home alone together as usual. His dad was at work, and his older sister Teri was at school, finishing up the last full week of classes before Christmas vacation.

As he often did in the early morning, Jeff had snuck out of his own bed and crept down to his parents' room to snuggle in their bed and fall back asleep.

That's where he was when he woke later that morning to see flames all around the bed.

"Apparently my mom's cigarette had started the curtains on fire, which quickly spread throughout the room," Ford recalled. "I screamed and my mom came running in and I dove off the bed into her arms."

At first she could only pace back and forth, helplessly chanting over and over, "Oh my God, what have I done?" while Jeff watched, frightened and confused. Finally she gathered her wits and carried her son down to the basement, told him to stay there, and shut the door.

"She went back upstairs, I'm assuming to put out the fire. A short time later I heard her screaming my name over and over. I started up the steps to get to her, but as I opened the basement door a huge cloud of smoke came rolling in on me and I had to go back down."

The basement was quickly filling with smoke. Jeff might've been too young to comprehend what had happened, but he knew one thing for sure: he had to get out.

The family had two Christmas trees that year: one upstairs and another in the basement. Jeff wedged himself between the wall and tree, scrambling up the decorated branches to reach the narrow window well, about seven feet off the floor.

He could still hear his mother screaming his name as the fire cut her off from getting back down the hallway to where she'd left him.

He slid the window open and started yelling as he struggled to climb from the tree to the window.

"The neighbors were already trying to get into the house and heard my screams and came running. They pulled me the rest of the way out, and I kept screaming to them 'please get my mom, I can hear her screaming.' They broke some windows in the front of the house and tried to enter, but the fire was way too hot and there was too much smoke for them to enter. She stopped screaming when the smoke overtook her lungs."

According to a *Saginaw News* article recounting the fire, a neighbor first saw smoke coming from the dwelling as she was sending her daughter back to school after lunch. She and another neighbor heard Jeff's cries, and they ran to the back of the house and pulled him out.

The woman told reporters that when she asked Jeff how he'd known what to do he told her his father had shown him.

The township fire chief later told police the fire was probably started by a cigarette left burning in an ashtray too near the bedroom curtains.

The family dog, a beagle named Lady, was found behind the closed door of Jeff's bedroom, where she slept each night.

Rescuers carried the dog's body out and laid it on hastily spread newspapers in the connecting garage, where one of the firemen told reporters, "It didn't have a chance."

Mrs. Ford was found in the hallway, dead of smoke inhalation, about 15 feet from a living room display of her archery trophies, all covered with thick black soot.

Ninety minutes after the fire alarm first sounded, the township police chief was boarding up the blackened windows of the family's home. As quickly as it had started, it was over.

"My dad never picked up a bow and arrow again," Jeff said.

Jac Ford dropped hours at work so he could spend more time with Jeff and Teri, making sure they knew they were loved and cared for and that their emotional wounds were healing.

"He became our mother and father for three years. If it wasn't for his love and strength, I don't know if I would have made it."

For the next two months, Jeff stayed with neighbors Art and Betty Trumble while the family's house was being rebuilt—the same woman who'd helped pull him from the fire. After that, Jeff continued to stay there during the day while his father was at work and sister Teri was at school, until he entered kindergarten that fall.

As kind and caring as she was, she couldn't replace what he'd lost or fill the void inside him.

"I missed Lady and my mom terribly. I was so lonely for them."

Across the road from the Trumbles' house were a large woods and a field of tall wild grasses—the kind of place a young boy could wander and lose himself for endless hours on a summer's day. It was the world he'd yearned for back in the days before the fire, when "outside" was an exciting, unknown place beyond the safety of his home's four walls.

But the safety of those walls had proved to be a cruel illusion, hadn't they? And "outside" wasn't imaginary now. It was real.

For the first time in Jeff's young life, the door to that outdoor world was wide open. And it was right across the street.

As much as she'd loved Jeff's mother and honored her memory, Betty Trumble also knew full well what a young boy needed. It was an unspoken agreement between them.

"Betty would let me sneak over there and explore," Jeff said. "I would spend hours alone exploring those woods, sneaking along trying to see wildlife.

"I remember being amazed as a squirrel climbed a tree next to me, eye to eye, only three feet away, sensing that I wouldn't harm it. I sat there a long time, and we watched each other. I thought it was so cool how it could climb so easily and fluidly."

He returned, day after day, sneaking pocketfuls of birdseed and cracked corn. Patiently he coaxed the creature closer, until hunger and curiosity finally overcame its fear.

"I would throw it out a little closer each time, and pretty soon I had my friend right next to me. But it took a little while before he would eat out of my hand and many more trips before I could hold and pet him."

Even as the squirrel learned to trust the boy, young Jeff had discovered a world where he, too, felt safe and secure, a place with no walls to trap him, where the risks and rules made sense if your eyes and ears were open.

It was the start of Jeff's lifelong bond with wildlife, but he had no way of knowing then where that simple act of kinship would someday lead.

As he grew older, his exploratory missions went farther afield.

Still, it didn't help him escape the nights, or the nightmares they brought.

"I'd wake up in a pile of sweat, continually hearing her screaming in pain but never taking her mind off me even as she was dying. That is true love," he said. "The nightmares never ended for me. Even at 44, although not often, I still have to relive that terrible day in my sleep. Amazing how clear that day is in my mind, when I can remember almost nothing else about being five."

The only place he didn't feel alone was when he was with his dad, or when he truly was alone, outdoors.

Less than a mile from the house was the Tittabawassee River with miles of deeply wooded, undeveloped land along its banks. The river became Jeff's personal, private playground.

What Jeff couldn't have known at the time was that within a few short years his "playground" would be ranked among the most polluted sites in the nation, when scientists announced that the river's sediments and floodplains were heavily contaminated with toxic chemicals that had been steadily released into the waters since 1897 by Dow Chemical manufacturing facilities in Midland, just a few miles upstream.

The worst of those was dioxin, a toxic by-product of manufactured products, including napalm and Agent Orange, notorious chemical weapons of the Vietnam War. To this day, despite extensive cleanup efforts by Dow and the state Department of Environmental Quality, dioxin contamination persists throughout the soil, sediments, plants, fish, and wildlife of the river and its floodplain.

But in 1973, Saginaw County residents were still blissfully unaware of the public health hazard running through their midst. For the Ford family, the Tittabawassee represented an unspoiled wilderness in which to hunt and fish.

By the time he was eight years old, Jeff was wandering off on solo hikes up to six hours at a time, each trek leading him just a little farther down the river. He learned to sneak along quietly enough not to alarm the wildlife around him.

That learned patience paid off with rare wildlife encounters. He once watched a newborn white-tailed fawn nursing from its mother, staying there motionless and undetected while the little fawn ran circles around the doe before the pair casually ambled away.

Slowly, the whole family was healing too. By this time, Jac Ford had met and married JoAnn, a single mother with two boys of her own. She was a strong, loving woman with room enough in her heart to take in and nurture another mother's children. They were a real family again.

Sometimes in the summer they'd spend an entire month up north with Jeff's Grandpa Charlie on Wiggins Lake near Gladwin. The old man would wake him up at 5:00 a.m. to go pike fishing, assigning a name to every fish they caught and entertaining Jeff and the neighborhood kids with his stories.

Jac Ford began taking the boys rabbit hunting too. They had a new beagle now, Mitsi, who loved to bawl hot on the trail of a rabbit. Jeff was fascinated by the chase: the way the dog could follow a scent, how the rabbit would always come around in a huge circle back to its original position. Jeff was learning a new side to nature now, one in which man wasn't just a passive observer—he was a natural part of it, a predator in his own right, as much at home in the woods as the prey he hunted.

He was 10 when he realized there were species capable of turning that equation around. It would be the start of his lifelong fascination with rare, large predators at the top of the food chain.

The family had gone to Montana for a monthlong vacation. The highlight was a stop at Glacier National Park.

Jac wanted to hike up one of the trails leading to Granite Park Chalet, one of the most famous backcountry lodges in the world. Built in 1914–15 by the Great Northern Railway to provide what was then considered luxurious accommodations for adventurous cross-country travelers, the massive rough-hewn log and stone chalet is perched among the evergreens with a panoramic view of the park's snowcapped peaks. Granite Park was the last of the backcountry chalets built by the railroad and one of only two that survive today. The lights are propane lanterns, the nearest water is a quarter mile away, and the only bathroom facilities are pit toilets. It's now listed as a National Historic Landmark and continues to shelter visitors to the national park wilderness.

The plan was for the family of six to hike up the four-mile Loop Trail to the lodge, an altitude gain of about 2,300 feet. As often happens on a hike, the family gradually stretched out into two groups walking at different paces. Jac, Jeff, and JoAnn's younger son Matt were up ahead while JoAnn, Teri, and older stepbrother Max brought up the rear.

They started with the morning sun and had been hiking a few hours when the first group stopped along the edge of the trail for a rest, sipping water from their canteens and waiting for the rest of the family to catch up.

Matt heard the screams first.

At first, Jac and Jeff just laughed at the younger boy, teasing him that he was letting all those ghost stories of animal attacks get the better of him. Everyone who'd been to Glacier knew that the wilderness around the Granite Chalet was the site where two women had been mauled and killed by grizzly bears on the same night back in 1967, at two tent sites miles apart. They were the first bear-related fatalities in the park's history.

The legendary "Night of the Grizzlies" forever changed wildlife management policies in our nation's public parks. Before the tragedy it was commonplace for park visitors to feed bears and other wild animals. Indeed, garbage was often left exposed and treats intentionally used to attract bears to cars and campgrounds for curious, camera-happy tourists with little respect for or understanding of bear behavior or how potentially lethal it can be when wild animals become too unafraid of human contact.

By 1975, of course, the attacks themselves had passed mostly into myth and legend, but awareness of the grizzlies was never far from park visitors' minds and their presence something no wise hiker took for granted.

A few seconds later they all heard it, and this time it was unmistakable: the screams of a woman coming from somewhere down the trail in the direction they'd just come.

All three took off running back with Jac in the lead. About half an hour earlier, they'd passed a section of narrow trail along some steep, jagged cliffs. Had one of the younger kids fallen?

As they rounded a sharp turn, they saw their answer: a large adult male grizzly standing directly in the path between them and the rest of their family. Jeff couldn't see his stepmother or siblings, but he could hear them, frantically banging the metal cooking gear they'd hastily grabbed off their packs.

Until the men's sudden arrival, the bear had simply been standing there, confused by the noise. Now, the startled bear bolted and turned—running away from the men and directly toward the terrified woman and younger children.

About 20 yards away, the bear suddenly veered off the trail to the left and slowed to a walk. He glanced back twice at the humans, then ambled up the forested mountainside.

Unfrozen now, Jac and the boys rushed to their family's side. Jeff's mother was swearing and crying hysterically at the same time, beating her fists against her husband's chest. She wanted the hike over, and she wanted it over now.

In fact, the entire family would've liked nothing better than to head back down to the parking lot, but by this time the chalet was a good deal closer than the car. They pressed on up the trail. An hour later they ran into two park rangers on horseback who urged them to get up to the lodge as quickly as they could.

Once inside the chalet, they prepared and ate a warm dinner in the rustic but well-stocked kitchen, then fell into their bunks, exhausted. But Jeff lay there awake, unable to stop thinking about the bear.

It wasn't fear that kept him sleepless. It was fascination and a growing sense of wonder and respect.

If the grizzly was such a ferocious predator, why were none of them harmed? Even as a young boy, Jeff knew the adult male could have easily taken out the entire family. They'd had no weapons and had unwittingly cornered the bear on his own turf. Why had the potentially deadly encounter turned out as well as it did?

The family was later informed that the grizzly had been tranquilized and was being relocated to a more remote area of the park. That relieved their fears, and the rest of their visit was uneventful. But Jeff's curiosity lingered. By the time they returned home, he was buying every book he could find on the subject, intent on understanding this powerful yet curiously peaceful predator.

His studies taught him that grizzly bears truly desire no contact with humans and don't typically regard them as prey. But habitat reduction and recreational encroachment on wilderness have pressed bears and other large mammals into an ever-shrinking fringe, where unwanted human-wildlife encounters are becoming more frequent and also more frequently result in bloodshed as the animals become used to humans and begin to see them—and their garbage—as a food source when opportunity or extreme hunger arises. Often, it's the bear that ends up paying the ultimate price, when officials deem it too great a risk to leave anywhere near humans.

The more he studied them, the more Jeff came to respect and admire these powerful yet seemingly benevolent beasts and to realize how little he'd truly understood them at the time of his family's encounter. Rather than a rare stroke of unbelievable good fortune, his family's "narrow escape" had actually been closer to nature's norm.

According to the Ursus International Conservation Institute, a nonprofit organization dedicated to the conservation of bears and their habitat, the North American grizzly is a brown bear significantly smaller than its cousin, the Kodiak. Unlike the more carnivorous Kodiak, the grizzly's diet is mostly comprised of grasses, berries, and tubers, with only a small portion made up of carrion, insects, fish, and the like. In the natural order of the grizzly's world, people are simply not food.[1]

The vast majority of bear "attacks" are in fact not predatory in nature. Typically, they are sudden encounters in which a bear has been startled by a human's unheard approach and the bear reacts defensively. Still, even a startled grizzly will often bluff charge, then lose interest and retreat if it no longer feels threatened.

The greater danger is one that should be avoidable. Opportunistic by nature, bears of any species will quickly develop a taste for human food and garbage if it's left accessible by careless people. It is probably the most common scenario for bear encounters gone bad.[2]

The other high-risk scenario is unintentionally triggering a female bear's protective instincts by getting too close to its young.

At the time of the encounter, Jeff couldn't imagine anything more terrifying than an adult male grizzly. He didn't realize until later how lucky they were: the same surprise encounter might have worked out very differently had the bear been a sow with young cubs.

More than anything else, the female grizzly is known for being a good mother. The grizzly's slow reproductive rate—the entire cycle, from mating through rearing of young, typically takes three to four years—has made the species fiercely protective of its rare, slow to mature, and therefore precious young.

Abandoning her young—even to save herself—is simply not an option for a female grizzly. That's just not what a good mother does.

And, of course, no one understood that better than Jeff himself.

While he didn't realize it at the time, that sense of kinship would become one of the dominant chords of Jeff Ford's psyche, culminating in a single-minded obsession with chasing a phantom in a Michigan swamp. Some might have seen it as eccentric. Only a few saw it for what it truly was: a unique kind of love affair between a motherless boy and a misplaced wolverine destined never to mate or have young.

February 24, 2004,
First Sighting

✦ ✦ ✦

The snowstorm had started around 2:30 p.m. on the previous day, dropping from the sky in thick white flakes throughout the night all across
the flat expanses of Michigan's Thumb. It was still coming down
steadily at 2:00 a.m. when Aaron Schenk stepped out of the Verona bar
where he played in a weekly pool league. There were few other cars out
this late, so the trackless rural road was barely visible through the
snowy haze in front of his headlights. It made for a long, slow drive
home. He was glad when he reached home and could shut the door behind him and head for bed.

By early morning the harsh winter storm had passed, laying down in
its wake a six-inch blanket of fresh snow across the open landscape—
the kind of snow that stretches out like thick, heavy dust across the
open fields then drifts up high at the fence lines and the occasional narrow woodlots between two farmers' fields.

Aaron Schenk woke to the sound of the phone ringing at 6:30 a.m.

"I know you're up. You coming out hunting today? Look out your
bedroom window to the south about 100 yards. You'll see me parked on
the road."

It was his brother, Ryan, sounding too eager and excited for so early
in the morning. Aaron shook himself awake, peering out the bedroom
window to see Ryan's truck down the road, just as he'd said.

"I've got a big track out here—something we've never seen before."

For the past year, there had been a rash of unconfirmed cougar
sightings around Bad Axe, but the Schenk brothers hadn't put much
stock in those reports. In Michigan, cougar sightings are more popular
than UFO sightings and just as elusive when it comes to hard evidence.

People were always asking the brothers to put their dogs on a sup-

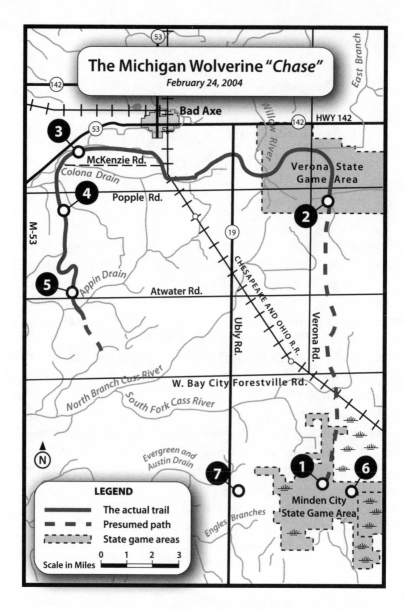

Map 2. The trail

1. The canals, the wolverine's sanctuary
2. Aaron Schenk backtracks the wolverine to this location
3. Aaron and Ryan Schenk first check out the wolverine's tracks
4. Ryan's dog Brandy catches the scent of the wolverine, and the "chase" begins
5. Harold Grifka's dog, Pluto, trees the wolverine
6. Jeff Ford's research site
7. Jeff Ford's residence

posed cougar, but they always rolled their eyes, more amused than irritated. There were plenty of coyote hunters in the county, and if there *was* a cougar in the area they'd have known about it sooner than the general population.

But that morning, driving home from his third-shift job at a small local machine shop, Ryan Schenk himself had spotted something unusual in the fresh white drifts along the rural roadway.

Aaron was skeptical but he knew better than to dismiss Ryan's suspicions. Quickly, he dressed and headed out to join him.

Standing there along McMillan Road next to his favorite hunting buddy, Aaron had to agree. The tracks were like nothing the two brothers had ever seen in all their years running their hounds after coyotes, foxes, and other game animals.

"It didn't really look like a cat track to us, but then we ended up trying it anyway," Aaron said later in recalling the day. "One good old hound he [Ryan] had was the only one that would even accept this track. Must be because it was something they didn't recognize or it scared them, but none of the dogs would have anything to do with it. But we put old Brandy on it, and she just went in. As she got close enough, the chase began."

They couldn't get a look at it, but whatever it was it refused to leave the cover of the one-mile section of woods where the dogs first roused it. It ran the dogs in circles, round and round in the woods, trying to lose them.

It was after 9:30 a.m. by then. The brothers were using two-way radios to communicate as they worked the track with the dog, circling around but never quite closing in. A pair of neighboring coyote hunters heard the radio chatter and joined in, intrigued.

Unable to resist the lure of this "mystery" hunt, Johnny Boland Sr. and Jim McKnight were soon part of the chase. Finally, Boland closed in enough to follow the dogs into the woods and try to get a look at their elusive prey. He radioed the other hunters from the woods, his astonishment clear even over the handheld radio.

"We're not running what you think we're running."

No further explanation came. Immediately, the questions began flooding in from Aaron and the other hunters on the same frequency. Boland still hesitated to reply.

The wolverine enters the research area with snow on its nose after a 12-inch blizzard.

"I would rather not say what this animal is on the radio."

The radio chatter was relentless now, everyone pressing him for more information.

Finally, Boland's voice broke out across the airwaves, loud and clear. "It's a wolverine."

The men laughed. No one had claimed to have seen a wolverine in Michigan in decades, maybe centuries. This was even more far-fetched than a cougar. John had to be dumb or crazy, they chided, or maybe a little of both.

The laughter turned into shouts of astonishment a short while later when the animal made its break and burst out into the open where all of them could see it: dark brown, long and low to the ground, the silhouette unlike anything the men had ever hunted before. There was no denying it now.

They called the Department of Natural Resources to alert it of their find. But the short-staffed field office was used to unsubstantiated claims of unusual animal sightings and simply lacked the manpower to chase down every call.

"The first call came in the morning. The woman who told the secretary she'd observed coyote hunters in the field near their house, and when she asked what they were chasing, they'd said a wolverine so she called the office," said Arnie Karr, a wildlife biologist now retired from the DNR who worked at the agency's Cass City field office at the time. "We told her that wolverines had been extinct in Michigan for over 200 years so probably it was something else. We didn't take it real seriously."

In fact, Karr initially thought it was a joke.

"That night was the big Michigan–Michigan State game so I told the secretary she'd probably have somebody else calling back soon saying coyote hunters were chasing Sparty, like somebody just being cute playing a trick on us," said Karr, chuckling. "But as time went by and we got a couple more calls, we got more curious."

Curiosity was an understatement for the local hunting community. By this time, the brothers' radio traffic was attracting attention all across the Thumb region. More and more people were tuning into their radio frequency, listening to the excited chatter flying back and forth as word spread of the find. More men and hounds joined in the hunt, with cars and trucks pouring into the area like the circus had come to town. Soon the snowy woods and farm fields were alive with the hunt.

"There were ample guys trying hounds on it, but it was just something our dogs had never smelled before, nothing like that, and some were scared," said Aaron Schenk. "I was putting dog after dog on the track. With most of them, the hair just stood up on their back and they wanted nothing to do with it. These are the same dogs that run bear with no fear, so them being scared of the tracks surprised me."

Even after the animal broke out of the wooded section where the track began, somehow it still eluded them, running south in haphazard fashion, crisscrossing five square-mile sections of woods and farmland with the massive hunting party in pursuit.

The animal darted along in a zigzag pattern, diving into the security of the next woodlot then bolting back out into the open again in a desperate bid to reach the next point of cover.

On two occasions during the chase, the wolverine allowed the dogs to get within a quarter mile, then it would suddenly kick into a higher gear and within minutes be a mile or more ahead again.

At one point the dogs got hung up in a woodlot, unable to pick up its trail. Minutes passed, and the dogs didn't reappear. Worried, the men went in after them, afraid the wolverine might have turned on the dogs and inflicted some damage.

They followed the dogs' route along the wolverine's tracks to a spot where they abruptly stopped at a heavy thicket. The wolverine's tracks had simply vanished! Confused, the hounds were circling aimlessly, trying to pick up the trail.

It took the men a while to figure out what had happened, and when they did, they almost couldn't believe it. Apparently, the wolverine had backtracked on its own path for nearly a half mile before jumping off to one side and heading east again.

A few miles later it repeated the same maneuver, once again leaving the dogs baffled and circling in its track.

"We couldn't believe this animal had the intelligence to trick both our dogs and us like that," said Schenk. "We had never experienced anything like it before. I mean, we hunt all over the state, and the athletic skills and intelligence [it] displayed that day was like nothing we had ever encountered before and will probably never experience again."

By noon, Ryan Schenk had to drop out of the hunt to get ready for another day at work. But before he left, the brothers made a vow.

"It's kind of an unwritten rule [that] when you're in charge of a hunt it's up to you if you're going to kill your game for the day or run it. Ryan said, 'You were second in charge on the hunt so it's your call now. Let's not take the wolverine's life. This is way too cool.' I agreed with him."

But would they even get a chance to honor that vow? The frustrated baying of the hounds sliced the frigid air, but none could close the gap.

"When we saw the animal cross a road in front of us, we just couldn't believe how well it could keep up on top of the snow compared to the dogs. It outran them four to one. This thing was made for the snow, like it was running on snowshoes."

Finally, shortly after 2:00 p.m., the wolverine appeared in the open again in a square-mile section just west of McMillan Road near the Cove Sanitation Landfill. Any further escape south was blocked by the cars lined up all along Atwater Road. In every direction but one, there was nothing but open field as far as the eye could see.

That's when one of the lead hounds, a dog owned by hunter Harold Grifka, pushed the wolverine hard enough to force it back into the woods, treeing it in a massive, mature poplar.

It had been nearly eight hours since the chase began.

"I was standing right underneath it, 25 to 30 feet up in the tree sitting in a crotch right on the stump end of it," said Schenk. "I was one of the first guys to witness it. Unfortunately, I didn't have a camera with me."

Emboldened, Schenk broke off a small tree branch blocking his view, moving in close so he could get a better look.

"When that branch snapped the critter peed and sprayed me," he said, laughing. "You wouldn't believe the stink. I washed and washed it but had to throw that sweatshirt away, it stunk so bad."

Once word spread that the wolverine was treed, a huge crowd of curiosity seekers swarmed in to join the hunters. By midafternoon, 30 or 40 cars and trucks were stretched out along both sides of the road as far as the eye could see. Everyone wanted a look and a picture.

But the animal at the eye of the storm was silent, unmoving.

"It was very, very calm once it settled in and we started taking pictures, which was crazy. Normally when you approach a feline or canine that's cornered, they don't get comfortable. But this wolverine just settled in up there, kept its eye on certain people here and there, watching its back. Even with all those camera flashes flashing at it, it sat right in that one spot, never got edgy or skipped around."

The hunters had tried earlier in the morning to get a local conservation officer on the scene but had had no luck. Maybe things were different now. Another of the hunters called the Cass City field office again.

"Understandably, I guess they were apprehensive of it just as everybody else was. They kind of blew it off at first. But once we had it in the tree, another of the guys called them back and said, 'This is going to go down in history so we'd sure like you to come out and confirm it.' And that's what they did."

Within a half hour of that second call, Arnie Karr was pulling up along the road in his unmarked car, where he was stopped by one of the men.

"He asked what I was doing there so I explained I worked for the DNR [Department of Natural Resources]. He said 'Are we doing something wrong?' and I said, 'No, not at all.' Then he told me they were looking at the wolverine. When I asked if they were sure it was a wolverine, he said, 'I guarantee it.'"

But Karr almost missed his own opportunity for a first-person sighting. Just moments earlier, the wolverine had been startled by the sudden arrival of another group of curiosity seekers on snowmobiles. By this time much of the crowd had dispersed; perhaps the animal finally sensed its long-awaited opportunity for escape. It bolted down from the tree, fleeing out of the woodlot and into the open field with the snowmobiles in hot pursuit.

That was Arnie Karr's first sight of the wolverine.

"I was still out at the road," he said. "It was running in an open field about a quarter mile away with a couple snowmobiles nearby. One of the snowmobilers called the guy with me and asked if we could see it from the road, so he told his buddy to come pick me up."

Karr climbed on, and seconds later he was scrambling to document the event on a small digital camera.

"As we were riding alongside and it was running, it looked like it was floating on air," said Karr. "It had such large feet, they stayed up on top of the snow like it was running on pavement. We were going maybe 15 mph and it struck me that this was an animal that could go like that forever."

One of Karr's last shots was of the wolverine racing off into the distance, leaving nothing but its footprints in the snowy foreground. The picture made the front page of the *Detroit Free Press* the next morning. Others were shooting video footage that would make its way to the evening news all across the nation.

"I even had a buddy in Arizona call me that night saying, 'You're on CNN right now.' It was amazing. I've never been part of anything like it," said Schenk. He wasn't as pleased with the snowmobilers but shrugged it off as something beyond anyone's control.

"They got some awesome pics but I've got to admit it aggravated us a little, what they did. We had already run her plenty that day and our hounds had to work for what they got. It just didn't seem right.

The wolverine might have been fair game for anyone after that. But Aaron Schenk didn't forget the vow he'd made with his brother.

"I'll admit there were people out at the road offering us money to let someone shoot it. But we did a great job with the honor system. We had a meeting out on the road, and I said if there's one thing was going to happen it's that nobody was shooting this wolverine. I was going to sit there and protect this animal as long as I could; that was our choice.

"Something like this, you just don't take. There were no laws against it. But I knew there'd never been one recorded here or at least not very many. I hunt all over the state and never had seen nothing like this. They respected me for it. You take from the bountiful, but you don't take from the not."

Karr immediately sent off an e-mail message and pictures to the DNR headquarters in Lansing. The Lansing staff took over and prepared an emergency order to be signed by DNR director Rebecca Humphries to protect the animal from any kind of harassment or hunting.

"The wolverine was not even listed as a species existing in Michigan at the time so there were no protections, nothing in the law," said Karr. "The Director's Order didn't make the wolverine an endangered species in Michigan, but it protected that particular animal."

Meanwhile, the other hunters followed through and honored the brothers' request. The hunt was officially over.

"That's how it got to go another day. I felt really proud of everyone for that," said Schenk. "The following morning at 7:00 a.m., the Natural Resources Commission passed an emergency [measure] to protect the wolverine."

By the time the DNR announced the protective order the next morning, Schenk was already back out in the field, retracing the wolverine's route across 12 miles of farm sections and woodlots in an effort to learn where it had come from. Finally he found himself 10 miles east and two miles south of the spot where Ryan had initially cut her track along Barrie Road.

From there he followed the trail back to where the animal had initially come out of the thick cover of the southeast corner of the Verona State Game Area.

The track leading into Verona came directly from the south where, 7.5 miles away, lay the northern fringe of the Minden City State Game Area and its series of deep canals through the swamp—close to the very spot where the wolverine's body would be discovered six years later.

He knew all the tracks had to have been laid sometime after the snowfall ended in the predawn hours of February 24. By the time the

Schenk brothers discovered the track later that morning, the wolverine had already covered nearly 20 miles! It was mind-boggling to think how much more ground the animal had been forced to cover by the time the chase ended that afternoon.

It made him even more glad the hunt had ended with the wolverine's escape.

But Schenk and his buddies soon found out the media doesn't call off its own kind of hunt quite so easily or fast.

"Hounds men just don't try drawing attention like this. A lot of people look down on what we do, and we sure didn't want that kind of attention. But there wasn't a lot we could do about it," said Schenk. "They interviewed all us hound guys who were included in the chase. It was hectic. It's good fox and coyote hunting that time of year, but we got very little hunting in there for a couple weeks.

"It was crazy. It was one of the highlight hunts of my life, but I really don't know if I'd want to go through all this again."

Karr found himself in the middle of the same media firestorm.

"By the next day," he said, "I was on the phone with the AP [Associated Press], local newspapers, people from all over. Over the next several days we had calls from all over the country.

"It was understandable because it was just so unusual. It still is. The only other documented incident anywhere close is a wolverine footprint observed a couple hundred miles north, near Thunder Bay on the north shore of Lake Superior, which is a long, long way away from the Thumb. That was maybe a year or two after this sighting. There's been no other since." [AUTHOR'S NOTE: According to Neil Dawson, Wildlife Assessment Program Leader for the Ontario Ministry of Natural Resources, this may be a reference to a wolverine trapped north of Thunder Bay in 2004. Tracks were observed in 2006 near Wakami Lake Provincial Park near Chapleau, Ontario, roughly 250 miles (400 km) north of the Thumb and about 410 miles (660 km) east of Thunder Bay. An Alaskan crew doing wolf surveys for OMNR research scientist Brent Patterson reported wolverine tracks south of Wakami on January 26, 2006. See the map on page 123.]

Schenk is still awed by it, even looking back years later to that snowy late winter day.

"The day of the actual chase, we didn't know it was the only one.

The wolverine stares to the west.

But the day after they passed the emergency bill, I got a call from a buddy in the Michigan Hunting Dog Federation, who told me they'd passed the bill and it was now protected as the only wolverine in Michigan's history in more than 100 years," said Schenk. "It gave us a good feeling of gratification. It was awesome. We knew we'd done something no one else had done."

February 24, 2004,
Later That Day

✦ ✦ ✦

Jeff Ford heard the news first from a clerk at a party store on the 20-minute drive home from teaching science classes at Deckerville High School. A wolverine had been run by a coyote hunter's dogs up near Ubly, she said, a small town less than a half hour from where they stood talking.

He was only slightly less skeptical than if she'd told him she'd spotted a UFO landing in a nearby farmer's field.

"She was very excited but I must admit my first reaction was she was full of it," said Ford, chuckling. "It was such an odd and rare disclosure. I tried to act excited as she and I talked about it, but I left not feeling at all that her assertions were true."

Still, he couldn't stop pondering the notion as he drove home. Sightings of rare and improbable animals—from cougars and Blanding's turtles to Mitchell's satyr butterflies—have long been a cherished pastime of Michigan's amateur naturalists.

Incidents involving large predators have always caused the most excitement and controversy, capturing the imagination of the general public—and sometimes, unfortunately, igniting their fears and prejudices. The history of Michigan's gray wolf population is a prime example.

Once found in all 83 of Michigan's counties, the gray wolf was seen as a serious threat to livestock and game, a problem that required an aggressive predator control program. From the late nineteenth to the early twentieth centuries, a state-paid bounty made the gray wolf a serious target for elimination.

It almost worked.

By 1840 the gray wolf could no longer be found in the most populated southern half of the state. By 1910 they had vanished entirely from the Lower Peninsula. By the time the bounty was repealed in 1960, sightings had become extremely rare even in remote regions of the Upper Peninsula, with the last known pups born in the mid-1950s in what is now the Pictured Rocks National Lakeshore.[1]

The state's attitude toward the gray wolf was officially reversed in 1965 when it was given full legal protection and then elevated to the status of endangered species under the Endangered Species Act of 1973. Instead of elimination, now the goal was to save them. But was it too little too late? And were Michigan residents ready to embrace an animal they'd viewed for so long as a nuisance and danger?

In 1974 the state tried unsuccessfully to relocate two Minnesota pairs to Marquette County in the Upper Peninsula. They were all killed within months. Any further efforts to artificially introduce a breeding population seemed doomed to failure. Through the 1980s, only an infrequent lone gray wolf was ever seen anywhere on Michigan's mainland, their only stronghold an isolated population at remote Isle Royale National Park in northern Lake Superior, nearer the borders of Canada and Minnesota than Michigan.

The gray wolf had become the gray ghost.

That might have been the end of the story, but the wolves themselves had apparently never heard the rumors of their demise. In 1989 the tracks of two wolves traveling together were confirmed in the Upper Peninsula. In 1991 the pair produced the first litter of pups documented on the mainland in 35 years.

Today Michigan's gray wolf population is numbered in the hundreds and is considered one of the state's biggest wildlife recovery success stories. Scientists believe the current population is descended from lone animals that migrated to the Upper Peninsula from populations in Wisconsin, Minnesota, or Ontario. They had returned not due to human intervention but through the absence of it. Once humans got out of the way, the wolves came back by their own means, and under their own terms.

Jeff Ford couldn't help but wonder: could the same thing be happening with wolverines?

He was savvy enough to realize public statements of that sort were bound to get him in trouble and mark him as belonging to a lunatic fringe.

Wolf sightings—no longer rare—now only fan flames of heated debates over just how widespread they are, their potential impact on livestock and game, and whether their status should be changed from protected species to game animal for legal hunting. [AUTHOR'S NOTE: On January 27, 2012, the U.S. Fish and Wildlife Service (USFWS) officially removed wolves in the western Great Lakes region from the federal endangered species list. Management authority over wolves in Michigan has been officially returned to the DNR, putting the state's Wolf Management Plan into effect. The goal is to control problem wolves while maintaining a self-sustaining population. New state laws also took effect on January 27, 2012, that established guidelines for legally killing a wolf under certain circumstances to protect livestock or pets. Currently, wolves remain a protected species in Michigan with no hunting or trapping season.]

More recently, a debate over cougars in Michigan had reached near-manic proportions in the past decade, with one side passionately proclaiming the existence of a remnant population while the other just as passionately debunked it all as fairy-tale wishful thinking most likely attributed to bad science and overfed family house cats.

The public battleground has been littered with wild speculation, scores of unsubstantiated sightings, and ugly accusations flying in both directions. Even tracks verified in remote parts of the Upper Peninsula have failed to quell the angry fires. They simply have added fuel to an argument now centered on whether the animals are escaped or released pets, naturally dispersing populations moving eastward from North and South Dakota, or a breeding population that had been undetected all along. [AUTHOR'S NOTE: In a September 13, 2011, press release, the DNR confirmed the presence of a cougar in Ontonagon County in the western Upper Peninsula. The animal's image was captured on a trail camera on private property on September 8, walking directly toward the front of the camera and clearly showing it has an ear tag and radio collar. According to the DNR, only western states currently have cougars collared for research projects so it is likely the animal traveled a great distance to reach the Upper Peninsula. As of this writing, no other details on the cougar's origins or status were available.]

Is it any surprise, then, that a lone wolverine—the rarest of the rare—might spark a similar flurry of attention?

Surely if it had really happened and an actual wolverine had been sighted, it would be on the news, right? Ford switched on the TV as soon as he got in the door, carelessly flipping channels until he found a local station. The first thing he saw was a promotional spot teasing a wolverine news story for the 6:00 p.m. news broadcast.

Two hours later he was planted in front of the set. Thankfully, he didn't have to wait for the newscasters to wade through their daily litany of car crashes, fires, and weather. The wolverine story was the evening's main headline.

He couldn't stop grinning at what he saw next: actual video taken by observers from their snowmobiles, running alongside and behind a long, low, brown animal bolting across the snowy fields.

Michigan's wolverine was no longer the stuff of myth.

It was real.

God damn.

At that same moment, countless thousands of Michigan residents were undoubtedly sitting in their own living rooms watching those same few seconds of televised footage. They'd probably talk about it at dinner that night or to coworkers at the office the next day. Some might even follow it up in their local newspaper or online for a day or two. Then they'd promptly forget it as just another piece of interesting but useless trivia.

Not Jeff Ford. He was already packing up his gear for a tracking expedition.

"That night I lay in bed wide awake most of the night thinking about the wolverine. By morning when I got up for school I had already decided I was going to find it," he said.

He'd already spent the prior evening online reading everything he could find about wolverines.

"My research indicated a wolverine has massively huge tracks as big as a bear. With all the snow on the ground, I figured I had a good chance of finding her."

But the next day, Mother Nature put an icy wrench in Ford's plans, dumping a fresh blanket of snow across the region. All the tracks from that first encounter would be buried in an endless expanse of white, rendering useless his original plan to pick up the trail where the hunters had left off. Still, he spent two days scouring the area for a sign.

"Because it was kicked up near a local landfill, the rumor was that it had come across from Canada on a garbage truck. So after two unsuccessful days of looking for tracks, I went directly to the landfill and talked to the folks there."

They pointed out the southeastern direction in which the animal had been heading and told him all they knew of the story. Ford knew there were two large tracts of public land in that general direction where it might seek refuge.

He spent a good part of that weekend and the following week searching nearly 900 acres of land in the Sanilac State Game Area, the first possibility along that suspected southeastern route. Nothing.

During the week at his teaching job at Deckerville High School, Jeff usually spent his lunch hours hunkered down with Steve Noble, the high school principal. Longtime hunting buddies, the pair could usually be found comparing field notes on deer-hunting tactics or fly-fishing techniques.

Nowadays, however, most of their conversations revolved around the Thumb wolverine and where it might have gone. It became their favorite lunchtime game.

"The topography of the Thumb is not real diverse. We're a lot of open farm country and small woodlots. Knowing the wolverine's characteristics, the kind of habitat they normally survive in is certainly not accommodated in the Thumb terrain," said Noble. "When we started looking on a map of the Thumb at places this animal was going to go, the only place that offered somewhat of a minihabitat was this piece that's seven square miles with not a road to it. We knew this was where she was going to go."

Noble was referring to the Minden City State Game Area—an 8,725-acre no-man's-land of bogs and brush, eschewed even by many hunters who prefer dry boots and less overgrown terrain.

It was the perfect place for a wolverine to hide.

Noble knew Minden well. He'd been hunting there with his dad since he was maybe 10 years old. Members of the Noble family hadn't owned any hunting property of their own so they'd always relied on public land. While it was just a short drive from the family home, Minden's inaccessibility made it almost as exclusive as a private sportsmen's club up north.

"Getting back to an area like that is physically demanding. Where we went was over an hour's walk, and most people just didn't go that far in," said Noble. "It was an area we kept to ourselves. We didn't invite a lot of people in."

But over the years, as Steve Noble grew into adulthood, it became increasingly difficult for the senior Noble to accompany his son across the punishing terrain.

"Finally Dad said you need to find someone you can trust to keep you company. That's not a place you want to go in on your own. You want someone else to be with you and know where you're at. So I extended that invitation to Jeff."

The pair began exploring Minden together in 2001, and it had become a favorite mutual stomping ground for everything from deer to rabbits.

"We'd been back in that area hunting for three or four years prior to the wolverine showing up in the Thumb, so we were both familiar with the habitat, knew the rigors of the area, knew a lot of people were not going to penetrate that deep."

After being tracked by hounds and run by snowmobiles, that wolverine would want to get as far away from people as possible, they figured. They also knew the area had a healthy population of snowshoe hares, a staple of the classic wolverine diet.

On March 6, 2004, Jeff Ford and Steve Noble launched their first foray into the Minden game area, plunging deep into the swamp in search of their prey.

Even though the Minden bog was over 10 miles from the area where the animal had initially been spotted, Ford knew a wolverine could travel tremendous distances. He felt certain it was a good possibility. That swamp was the most secluded place in the Thumb.

It didn't take long for them to realize it was also a very large haystack in which to search for one lone furry needle.

For most of that day, they just wandered aimlessly, looking for any kind of evidence that the wolverine might be there.

"We followed rabbit tracks, hoping to find wolverine tracks amongst them. But we weren't finding anything," said Noble. Still, they made one last try, striking in from the east side.

They were in their third hour of looking and 1.6 miles from the truck when they came across a set of unusually large tracks.

"They weren't fresh. They were old and deteriorated. But just based on the dimensions we knew they had to be from a bear, wolverine, or cougar—but just not enough specifics to determine which one," said Jeff.

In reality, none of those three species was at all likely to be found in this part of Michigan, but a bear was the most reasonable. A black bear had been confirmed nearby a couple of years earlier. While they hoped it was the wolverine, they knew better than to jump to hasty conclusions. They snapped a few digital photos but decided to keep the tracks to themselves until they had something more solid.

"At this point we knew people might think we had lost our minds," joked Jeff, chuckling.

The next weekend Jeff went back alone on another scouting mission, this time going in from the south side. He was more than a mile in, checking the thick brush along a series of drainage canals near an old scout camp, when he came across another set of large tracks.

"These tracks were made within the last week because we'd had a pretty good dumping that Tuesday and Wednesday. But it doesn't take very long for tracks to deteriorate in the snow, and I was still unable to count the number of toes and toe placement that were critical for positive identification."

Jeff refused to get discouraged as he continued dogging the trail. A short while later the persistence paid off with his first solid clue that the animal he was trailing might indeed be the wolverine.

The tracks led to a spot with a series of branches slightly more than a foot off the ground.

"Whatever it was, that animal went under those branches with no drag marks from the underbody. I knew then it could not have been a bear, and I highly doubted a cougar. I immediately called Steve on my cell and told him this is probably the wolverine."

But more days passed without another sign. By then an early spring thaw had begun, any chance of tracks vanishing with the melting snow. He had to focus on areas where it was muddy but not submerged, often forced to move along the beaten path of deer runs to make his way through the heavy brush.

Long weeks continued to drag by without a sign.

Finally, on June 2, 2004, he was hiking about 1.5 miles in along the north side of the bog, moving along a well-established deer run, when he came across a huge track in the soft mud.

"The clarity and dimensions of the track were awesome. I started to follow the run and soon found hundreds of these same tracks stretching off maybe a half mile."

The next day at school, he and Steve compared them to copies of a wolverine track printed in a book of North American animal tracks written by well-known Michigan trophy hunter Richard P. Smith.

Wolverines have five toes, like a Michigan black bear, with a track that's similar in size: about five inches long and four inches wide. But while a bear's toes are all out in front of the pad, a wolverine has a larger thumb on one side and a small "pinky" on the other, leaving three toes pointing forward. With a clean impression, there was no mistaking the distinction.

The muddy tracks were a perfect match for Smith's example of a wolverine.

Ford placed his own hand near one of the prints, so eerily shaped like his own human hand—minus, of course, the one-inch claws sticking out in front.

"As I followed along behind those tracks, I had goose bumps knowing I was following such a rare and unique mammal, one of the rarest in North America," he said.

The day after that, he and Noble returned with a video recorder, casting plaster, and water and set about documenting their find. Noble came with him on the third day to make even more casts, and they followed the trail for nearly a half mile, recording it on a small video camera.

Then they called DNR wildlife biologist Arnie Karr, who had made the initial confirmation in 2004, and arranged to bring him the casts to confirm that the wolverine was indeed alive and well and still inhabiting the Thumb.

"I wasn't really surprised. It wasn't all that far from the initial chase. That was in south Huron County, and this track was in north Sanilac County right next door," said Karr." There's a high number of roadkill and dead deer in the Thumb, and the Minden City State Game Area has a good population of rabbits and rodents, too, so we thought it'd have plenty of food."

Still, it was exciting news even for staffers who routinely studied wildlife for a living. No matter how it got there or when it had arrived, there was no denying now that Michigan did indeed have its own resident wolverine.

Now that he was sure the wolverine was actually living in the area and hadn't simply passed through on its way to somewhere else, Ford and Noble devised a new game plan. Over the next several weeks, they would take Jeff's old trail camera—the film kind with a shutter that's tripped by motion—and begin randomly positioning it on tree trunks along the deer run where they'd found the initial tracks, in hopes of getting a lucky shot.

The DNR didn't try to stop them even when they later learned they were setting camera lines.

"We felt the wolverine was an oddity that likely involved some kind of human intervention that got it there in the first place, so our feeling was to just leave it alone. But we really didn't see any harm in what he was doing," Karr said after the first photos came out.

For a solid month, the camera faithfully recorded every critter that wandered past its lens: deer, coyotes, rabbits . . . and literally hundreds of pictures of weeds and branches blowing in the wind.

But Jeff refused to give up hope or abandon the mission. Dutifully, he trudged back out through the thick underbrush every week to 10 days to check the camera, often accompanied by Steve but just as often alone.

"Initially we were doing a lot of the stuff together, but I coach two different sports and our schedules wouldn't always match up, so Jeff would go in and maintain the cameras himself sometimes," said Noble.

But while the demands of work and family were making Noble's forays less frequent, Jeff's single-minded obsession had only begun.

"I began laying my camera in various locations trying to catch him strolling by, but nine months later I had yet to get a glimpse of the beast," said Ford. "But he continued to tease me, torment me if you will, by continually leaving his calling card. Those tracks were so big, they stood out like an elephant in a swimming pool."

But week after endless week the trail camera noted nothing but the wind.

By September of 2004, Jeff had nearly 500 pictures of weeds and branches randomly blowing in the breeze. The girls at the drug store's one-hour photo-processing booth began to tease him good-naturedly, calling him "the Brush Guy" whenever he showed up with another roll of film in hand.

Fall was giving ground now as another relentless, cold winter pushed its way into the Thumb. Deer-hunting season was fast approaching. Now the school lunch hour conversation between the hunting buddies returned to tales of deer stands, food plots, and the state's game management plans.

Even Jeff had to admit it was a welcome change in what had become a monotonous, uneventful routine. They both knew there would be little chance of spotting a wary animal anyway with deer hunters moving through the woods. He was also leery about leaving expensive equipment out where anyone might stumble across it. The camera went back home for safekeeping, replaced with a hunting bow as Jeff turned his focus to a goal with more obvious rewards: stocking the family freezer with fresh venison.

But the wolverine was never far from his thoughts. By January, he was ready to start the search in earnest again, and there was enough fresh snow on the ground for good tracking.

Steve was ready to get back at it too. Their plan was to first confirm that the wolverine was still in the area, then position the trail camera in strategically selected spots in heavily frequented locations.

But that turned out to be trickier than they'd hoped. Like a ghost, the wolverine had vanished once more, leaving no trace in the snowy swamp. They searched for hours, tracing the zigzagging trails of rabbits bounding across the drifts and the familiar cloven tracks of deer. The signs of the swamp's many inhabitants were easy to spot: the rounded impressions beneath low-hanging trees where a group of does had bedded for the night, buck rubs deep in the tamaracks, or young bushes gnawed clean as scissors by a hare's sharp teeth. But of the wolverine there was nothing. Not a trace.

Jeff decided to change strategies when he returned a week later on his own. This time he'd strike in from the south end off the dead end of Reinelt Road near the Boy Scout camps, within a mile of the canals where they'd found tracks the previous June.

The canals would be a likely spot for the wolverine to hang out in, he reasoned, providing a safe haven just as they had when it escaped from the coyote hunters the previous spring. He was excited at the thought of it as he hiked in, focused on the mission ahead.

What he forgot to pay attention to was a recent warm spell that had hit the area and its impact on the frozen water beneath his boots.

He heard a loud crack, and suddenly the world was dropping out from under him. The only thing that kept him from going under completely was the bottom of his elbows hitting the ice.

The wet and cold rushed through his heavy clothes like an icy freight train. He knew he had to get out of the water fast. He clawed at the edge of the ice, struggling to climb back out on top. But each time he lunged forward and grabbed the sheet of ice, the weight of his body would break and tip another section down with him, submerging it. Ice-cold water ran over his hands and arms. He was now soaked through to his shoulders, only his head and upper back still dry.

He knew he was in serious trouble now, that he'd begin losing dexterity and strength within minutes. And there would be no help coming. If he was going to get out of this, it was up to him and him alone.

With one huge effort, he heaved himself forward again. This time he managed to grab hold of the sides of a huge slab and pull himself on top of it. The weight of his body submerged it slightly, but it remained fairly level, providing enough support to allow him to crawl onto the solid icy bank.

He might have been out of the water, but he wasn't safely out of the woods, not by a long shot. It was 18 degrees Fahrenheit, and he was a good mile from his truck, soaked through and half frozen, mentally and physically shaken and disoriented from the shock of the experience.

The Global Positioning System (GPS) unit around his neck was useless—it had been completely submerged in the water and was dead. All thought of tracking the wolverine was wiped from his mind. Now his survival depended on following his own tracks backward in the snow.

A quarter mile into the trek, his pants were cracking with each step, the wet fabric turning to ice. Another quarter mile and the skin and flesh of his extremities were completely numb, his feet encased in ice water inside his boots.

His only chance was to get back to the truck as fast as was humanly possible. He kept going, trudging along in mindless urgency. There it was at last!

He got in and turned the key. As the engine roared to life and the heater slowly kicked in, he stripped off the wet clothes. The only thing he had that was dry was a shirt in the back of the crew cab. He was shaking so violently that he could barely get it on over his arms. Half-naked, he climbed behind the wheel and started for home. He could barely drive at first he was shaking so hard. His hands and legs were not working right, and his private parts had virtually vanished.

About 10 minutes into the drive, once the heat was flowing through the truck and the pain and violent tremors had stopped, he was able to start laughing at his own stupidity. What a ridiculous, frightful sight he must be! As desperately as he wanted to get home, he forced himself to drive slowly, obeying every traffic law. The last thing he wanted was to have to explain to a police officer why he was driving around the countryside half-naked in the middle of winter. He could see the headlines now: "Local High School Science Teacher Suspected of Being a Pervert."

It wasn't until February that Jeff and Steve again found telltale tracks, about 1.5 miles from the nearest road in the gut of the swamp, not too far from where they'd found those first indistinct tracks the prior winter.

It was a Saturday afternoon, cold and dreary. The two men had already hiked over five miles in the Minden City swamp and were exhausted and frustrated over their lack of success.

A tired, empty silence had fallen between them. Jeff was lost in his thoughts, depressed by his growing suspicion that the wolverine was either dead or gone for good. They made one last huge sweep and were coming around in a circle, headed back to the truck. Jeff wasn't even scanning the ground with his eyes anymore. He'd given up.

In his mind, he spun out an endless array of horrible, negative scenarios for what might have happened.

The wolverine had been shot by some hunter during shotgun season. Maybe it'd come too close to his ground blind and he'd freaked out and opened fire on it.

Or maybe some jerk in a tree stand had seen her and taken the shot, wanting its head and pelt for his trophy wall. Jeff could feel his anger rising at the imaginary poacher, gloating over being the only person in Michigan to kill a wolverine.

Then that thought slid away, replaced by the image of the animal crossing Mills Road at the southern edge of the swamp, being struck by a speeding truck, and dragging itself back into the thick undergrowth to die.

No, it must have been trappers who killed it! Angry fantasies of revenge bloomed red as his imagination ran amok. He'd track them down and shoot them, tie their feet to a boat anchor and shove them beneath the ice . . .

Steve's voice jerked him back to reality.

"Look what we have here . . ."

It was a fresh wolverine track, unmistakable, the toes and claws clearly outlined and distinct. Even more exciting, it appeared to be no older than that morning or the previous night. They were close!

They followed the trail to a second set of slightly older tracks, indicating that there was some habitual pattern to its movements.

Judging by the pattern of tracks from the past year and the current winter, the wolverine appeared to be living in a roughly half-mile area known to locals as "the canals" and roaming eastward at night to hunt the snowshoe hares that abounded in that region. Their hopes of getting the animal on film skyrocketed.

But now they changed camera strategies. Instead of just placing the camera randomly along game trails and hoping for a passing glimpse, they would bring the wolverine to the camera.

"We knew the wolverine was very smart and if we were going to get a picture, we would have to exploit the one weakness it had, and that's its belly," said Noble. "Wolverines are tremendous eaters. We knew we could get this thing coming in to some meat."

Jeff was in full agreement. In every scientific case study he'd read, the researchers had used some sort of carrion as bait to lure the wolverines to their cameras.

In Idaho, one highly respected researcher had rigged up a contraption with animal flesh and bone hooked to a cable to entice wolverines into his live traps. When the wolverine climbed in and tugged on the bait, the cable brought down the door.

In Alaska, another well-known researcher was using dead animals stuck in an upside-down bucket hung from a cable between two trees. The setup forced the wolverine to climb the tree and travel down a walking pole before it could look up into the pail and tug out the meat—creating a perfect pose for the camera to snap pictures of the animal's fully exposed front, where its fur displayed a distinguishing chest pattern unique to each animal.

In Glacier National Park, researchers strapped carrion to the top of eight-foot telephone poles driven vertically into the ground, which were then covered with wire brushes. As the wolverine climbed the pole, the brushes would catch hair for DNA analysis.

Jeff realized that their current methods had been doomed to failure from the beginning. No self-respecting wolverine was going to amble past some random trailside camera on a casual stroll through the swamp. They had to provide an incentive for the wolverine to come to them.

There couldn't be a better time to bait. In spring or summer the wooded wetlands would be bursting with abundant game, a feast of plenty for a cunning wolverine. But it was the dead of winter now, when the Minden swamp was a wildlife ghost town, with only the rare snowshoe hare or deer track to mark any significant sign of life. Some carefully placed carrion would look pretty tempting right about now to a hungry wolverine.

Jeff had some elk left in the freezer from a 2002 hunting trip to northern Michigan. He also still had a good share of venison from an eight-point buck he'd shot during bow season that fall. He volunteered to hump it all in as soon as he had another free day.

Jeff didn't care if Steve came with him or not. He knew he was getting close to finally getting that picture, and he said so to everybody who cared to listen.

It was his dad who gave him the biggest encouragement and confidence, just as he had when Jeff was a kid playing football.

"If there is anybody who could get a picture of wild wolverine living deep in a swamp, it would be you, son," Jac Ford told his son. "I look forward to seeing your first picture!"

As it turned out, Jeff didn't have to wait even until the next weekend to put the plan in action. A blizzard blew in across the Thumb that week, burying the countryside and canceling school for three days.

Since both he and his wife, Amy, were teachers, "snow days" were always a special treat—an unexpected day off together when they could snuggle back under the covers to relax and sleep in.

But this time when the call came at 6:00 a.m., Jeff wordlessly jumped out from under the covers and began to hurriedly pull on his hunting clothes and pack boots in the dark.

He heard Amy's voice call out groggily from the still warm bed.

"What in the world are you doing?"

He told her he was heading out to the swamp with a game camera and a 12-pound load of deer and elk meat on his back.

She just laughed.

"Yeah, right, in this storm. What are you *really* doing?"

He turned on the light, searching for his favorite face mask. Now she could see he was serious.

"You're crazy," she mumbled, rolling back over. "Don't wake Riley on the way out."

It was a miserable trip in bitter cold, with 40 mph wind gusts and almost zero visibility. Jeff slogged through knee-high snowdrifts, slowly pushing his way back to the spot he'd predetermined as the best location for the camera. The hike—normally 45 minutes—took more than two hours.

It took four hours to set up the camera as planned and to lay out the meat in a careful display about six feet in front of it. His legs were rubber by the time he was ready to head back home. But his spirits were high as those of a kid on Christmas Eve.

"I saw no tracks, but I wasn't too concerned because they would have been covered up in less than an hour in that blizzard," he said.

By this time, another hunting buddy, Jason Rosser, had started asking to join Jeff and Steve and be part of the quest. Jason's dad and Jeff's father-in-law owned hunting property together, so the two younger men had known each other for years. For months Jeff had been texting him about their efforts, sprinkling wolverine stories into their deer-hunting conversations. Rosser was hooked.

"It was mostly for the unique experience that I wanted to get involved," said Rosser. "It was something nobody's ever found in Michigan, just something really unique and interesting."

Noble welcomed the extra help too.

"We'd known right off the bat it wasn't going to be easy. The wolverine is a smart animal, and trappers have passed along a lot of history of putting out carcasses to bait them and the wolverine not coming in until coyotes had been there inspecting it first," said Noble. "But it was still very frustrating for a long time, setting those cameras up and not getting anything. It was enough work just going in."

Working as a team made it easier to keep forging ahead, he said.

"It was something we were all doing together, even though our schedules wouldn't always match. The goal wasn't to see who was going to lay eyes on a picture first. Our number-one goal was to get the picture. Who got it or got to see it first didn't really matter."

Since Steve had already told Jeff he'd be busy the next weekend and unable to go out, Jeff e-mailed Jason to give him an update and invite him along.

"I feel pretty good about getting a picture of him. If I do get a picture, it would be only the second confirmed sighting of a wolverine since before Michigan became a state, so I'm pretty fired up about it," he wrote in an e-mail to Jason on March 2, 2005. "I'm going back in a week to get the film and take some carrion out to him with new film."

Jason was just as excited. He, too, felt they were getting close, and now this was his chance to be a part of it. He wrote this reply on Thursday, March 3, 2005.

> If you don't mind I would like to go with you to the swamp next weekend if you're going out? If nothing else you can take my digital camera to put out. It is easy, you can't mess it up. You could write a pretty good story for the paper if you get photos. I have approximately 30 pounds of ground venison I'm not going to use if you want it.

The two men made the hike in on Sunday, March 7. As they approached Jeff's camera, they could see there were wolverine tracks circling all around it, within five feet of the camera on all four sides.

The venison was gone.

Jeff hastily swapped rolls of film and repowered his camera while Jason set up the new camera. They rebaited the area and headed back out as fast as they could. They drove straight to the one-hour photo counter at the Rite Aid drug store in Bad Axe.

From left to right: Jason Rosser, Steve Noble, and Jeff Ford.

"That was probably the longest hour of my life, waiting to see those pictures," said Jeff. "But finally they were ready. We tore the outer envelope off the packet and began furiously scrambling through the pictures."

Nothing. Not a single picture of the wolverine. What had possibly gone wrong? The answer turned out to be stunningly simple and obvious, once they thought it through.

"Having my camera on low power mode proved to be the mistake. Although low power mode preserves battery life, it takes a few seconds to power up when it senses motion. So by the time it took the picture, the wolverine was no longer in front of the lens."

But he wasn't about to give up, especially not now.

"It was an awful blow, but at the same time coming so close fueled my fire that it might really happen," he said. "I might be able to get that picture if I just kept trying."

The near miss was enough to whet Jason's appetite, too, and now both men were in hot pursuit. Despite all their efforts, their elusive prey still somehow managed to escape them. That next week they could scarcely think about anything except trying again.

Jeff wrote this e-mail to Jason on Monday, March 7.

I woke up in the middle of the night thinking about how close we came to getting him. If we don't get a picture by this weekend, we'll definitely have to change strategies. If he doesn't come in to bait by two weeks, it probably won't happen cause he knows it's there, and yours is close enough he will quickly know it's there too. Hopefully those big juicy chunks of turkey will be too good to turn down. There are two wolverines at a zoo near Frankenmuth. They feed them whole chickens so I think we're headed in the right direction.

If we get a little snow this week, that would be good. We should be able to tell if his tracks have been in on your bait. If he pulled the same stunt as mine, we'd be better off setting NO cameras over the bait, and putting all three in the perimeter. If he avoids the area altogether, we'll have to abandon the baiting, find his new entry to the area where the snowshoes are at, and set up the cameras as well hidden as possible.

I've decided to leave my camera out there 'til July and keep trying to get him, even if we get pictures in the next couple weeks. This is a once in a lifetime opportunity so we should take advantage of it. He may be gone next year.

If we do get a picture next weekend, the next time in before the snow melts we should track him to the canals and try to find his den and food cache. (But I have a funny feeling getting his picture is going to not be easy!) His home base should have plenty of tracks and sign. Let me know which day would be better. I would prefer Saturday, just because I'll be chomping at the bit to get back out there, but Sunday will do if we have to.

Jason's reply came at 7:36 that evening.

Jeff, I was thinking the same. I don't think I can wait till Sunday so Saturday will be just fine. I have to work till 1:00 so I should be at your house shortly before 2:00. I was telling dad about our findings and that we were keeping it low key for now, [and] he would like to go with us Saturday just to go for a walk because he has never seen that area. This is going to be one of the longest weeks ever for us but it's going to happen maybe not this week but we will get him. I agree we should leave our cameras out as long as we can and track him in deeper if we get some good snow. I have a full chicken that is not going to be eaten so I will thaw it a few days early and bring it with me. I talked with my sister-in-law and she thinks we have to buy the scraps but I don't think she knows what she is talking about.

I hope my other camera will be back by this weekend. I will bring some more meat and the chicken. If we could get his habits down eventually over time we might be able to put together quite a photo collection. I can't wait.

Saturday, March 12, was the next day they were both off work. They met early that morning and headed into the swamp.

They watched intently for tracks in the snow on the way in. It had been a few days since the last snowfall, so the wolverine would have had time to lay down plenty of evidence that it was still around. Both men felt confident of it.

But as they approached the cameras, they could see no sign of tracks anywhere. Jeff felt the air go out of his chest as his spirits sank in disappointment. He changed film rolls, happy at least to see all 24 shots had been used. If nothing else, the camera was still working. Jason took the memory card out of his digital camera, and the pair parted ways at the road, Jason planning to check his on a computer and Jeff headed back once again to the Rite Aid photo counter.

This time he waited out the hour by driving to a local bar and tossing back a couple of quick beers to calm his nerves. While he was there, Jason called to tell him the bad news: his camera had come up empty.

Disheartened but unsurprised, Jeff drove back to the Rite Aid store and walked up to the counter. The girl smiled and shook her head as she handed him the packet of processed shots.

"Well, I'm not charging you full price because you have 22 pictures of brush again, Jeff," she said, laughing.

She was still grinning as he reached for his wallet to pay.

"But that wolverine sure is pretty."

Ford's head jerked up, mouth falling open.

"What did you say?"

After 371 days of trying to photograph the wolverine and hundreds of pictures of nothing but brush, there it was in its splendor!

"That wolverine you took a picture of. She sure is pretty."

He threw a 10-dollar bill on the counter, grabbed the pictures, and thumbed through the edges as fast as he could.

"It was the last picture—the oo negative. It had the most beautiful picture of a wolverine I had ever seen. She was caught broadside, only six feet away, with her right paw up in the air and turned toward the camera, like she was waving as she strolled on by."

He marched out of Rite Aid with both arms held high like a winning quarterback exiting the field, pumping his fists in the air as he let out a single, triumphant shout.

He had found his wolverine.

March 14, 2005

✦ ✦ ✦

Simply getting a photograph of Michigan's only known wild wolverine would have been enough for most men. Not for Jeff Ford.

Jeff had already written one article about finding the wolverine's tracks in the September 2004 issue of *Woods-N-Water News*, a statewide magazine for hunters and other outdoors enthusiasts. Being able to provide a follow-up story with a fully documented picture would be even better.

But Jeff's real motives ran much deeper than that. Just as he'd done as a young boy when he'd read everything he could find in order to understand the seemingly inexplicable behavior of the Montana grizzly his family had encountered so many years ago, Jeff wanted to solve the mysteries of Michigan's wolverine. Was it a male or female? Where had it come from? How did it get there?

Back in 2004, when the wolverine was first spotted near Ubly in southern Huron County, the original hypothesis put forward by the Michigan DNR was that the wolverine had come across the border at Port Huron's Blue Water Bridge from Ontario, Canada, as a stowaway passenger on a garbage truck.

No one really thought it likely that the wolverine had been born in the state. Michigan had no viable wolverine population, had not even had a documented sighting in recent history. To have one suddenly show up in the Thumb by natural means was a stretch by anyone's measure.

"It was not only unusual to find a wolverine in Michigan, but the particular place in Michigan was so out of place," said retired DNR wildlife biologist Arnie Karr. "The Thumb's farmland is very different from the normal habitat that wolverines require to survive, and so far away from what is normal habitat for them, which is basically up around the Arctic Circle with a lot of brush, cedars, swamp—and snow and ice that lingers on into summer."

Wolverines mainly feed on carrion and cache their food in snow, said Karr, often using their powerful claws to dig down into the snow and ice to find dead animals. That frigid environment is a key element to their continued survival, he said.

According to the US Fish and Wildlife Service, the species was extirpated from the lower 48 states during the early twentieth century. Reestablished populations have moved down from Canada into the North Cascades Range of Washington and the Northern Rocky Mountains of Montana, Idaho, and Wyoming. South of the Canadian border, wolverines are restricted to areas in high mountains near the tree line, where conditions are cold year-round and snow cover persists well into the month of May.[1]

Michigan's Thumb might be cold by some standards—but it was hardly the ideal environment for a natural, spontaneous eruption of wolverines.

Since the animal had originally shown up in the vicinity of a Huron County landfill, was it possible it had hopped a ride there from Canada?

The idea made perfect sense, especially at a time when Canadian trash had become a hot-button issue in Lansing, the state capital. Over the past two decades, a series of seemingly well-intentioned but insidious events had slowly but surely turned Michigan into Canada's trash can.

The first occurred in 1978, when the US Supreme Court ruled that waste is an "article of interstate commerce," effectively limiting a state's right to regulate trash crossing its borders.[2]

That was followed in 1979 by the US Environmental Protection Agency's new waste management regulations, which largely replaced local municipal dumps with huge, highly regulated landfills.

In 1986, the Transboundary Movement of Hazardous Waste treaty between Canada and the United States officially opened Michigan's doors to Canadian trash. From there it was only a short step to making the argument that foreign waste was protected by the North American Free Trade Agreement (NAFTA) of 1994—a conceptual wedge that propped that door wide open.

On January 1, 2003, the city of Toronto began to ship 100 percent of its garbage to the landfills of Michigan.

Following years of nonstop public pressure, state legislators finally hammered out a "cease and desist" agreement with Canada, and the practice officially ended on December 31, 2010, when the last load of Toronto trash made its way across the border. But from October 1, 2003, through September 30, 2004—the time period when the wolverine likely arrived—11,558,899 cubic yards of Canadian waste was dumped into Michigan landfills according to the annual waste management report of the Michigan Department of Environmental Quality.[3]

All that waste didn't just magically appear in Michigan—it was trucked in, seven days a week, 24 hours a day. In 2004, the number of Canadian trash trucks crossing the border into Michigan skyrocketed to approximately 415 per day—more than double the prior year's estimated daily tally of 180 according to a Department of Homeland Security survey cited in a November 15, 2004, press release from the office of US senator Carl Levin.[4]

There was more than enough room for one little hitchhiking wolverine, right?

Wrong, as it turned out.

"We ruled out the garbage truck theory in part because we learned it all comes from Toronto," said Karr. "It would be as unusual to have a wolverine discovered in Toronto or southern Ontario as it was here. Also, they pack those trucks so full of trash nothing could survive the trash compactor."

The best guess, Karr said, was that it had escaped captivity—but from where or when there wasn't a clue.

"It probably escaped from some exhibit, or maybe someone had it in captivity and it escaped. People can raise them, but you have to get the appropriate permitting," said Karr. "Or quite possibly they acquired the animal illegally. There is an active underground network of exotic wildlife in Michigan. If you've got enough money, you can get just about anything."

Others weren't so sure. For some, that answer was just too neat and tidy—an easy way to blow the whole thing off. The DNR's explanation meant the wolverine wasn't a miracle of nature worthy of scientific study—it was just another stray pet, a mildly interesting novelty and nothing more.

"It makes sense, if you think about it from the DNR's perspective," said Steve Noble. "If we're confirming a natural progression of wolverines into the area, we now have to educate and manage a population. If

I'm telling you someone else brought it here, I'm able to dodge that bullet and not have to worry about trying to manage a species down the road."

That argument came as no surprise to Karr and other DNR staffers. In fact, it was a painfully familiar jab that had been stuck in the department's rib cage for years during the heated and highly public controversy over the status of wild cougars in Michigan.

According to the DNR, cougars were originally native to Michigan but were believed to have been extirpated in the early twentieth century, with the last known wild cougar taken in 1906 near Newberry in the Upper Peninsula.

In the century since, random reports of cougar sightings infrequently cropped up in various parts of the state, but none were fully documented with solid physical evidence and no definitive scientific conclusions had ever been drawn. So the continuing presence of cougars in Michigan wasn't exactly a hot topic of conversation among state wildlife biologists, and the notion barely even registered a random blip on the public radar screen.

That all changed with the arrival on the scene of the Michigan Wildlife Conservancy (MWC). Originally founded in 1982 as the Michigan Wildlife Habitat Foundation, the nonprofit organization's stated mission was wildlife habitat restoration. Gradually, however, the group's attention had turned more and more to the subject of cougars, led by its executive director, Dennis Fijalkowski, and Dr. Patrick Rusz, MWC's director of wildlife programs.[5]

In 2001, the group published a report summarizing all the purported cougar sightings and evidence that had appeared in public records and the popular press over the past century. Simply making a solid case for Michigan cougars was intriguing in itself. But Fijalkowski and Rusz didn't stop there. Instead they used their summarized findings to propose a much more far-reaching and controversial conclusion: that the DNR was wrong and a small remnant population of native cougars had been living and breeding in Michigan all along.

The DNR, for its part, largely dismissed the issue for lack of hard scientific evidence. If cougars were indeed in the state, they were likely to be escaped or intentionally released pets or random wild dispersers that had wandered in from parts of North America with established breeding populations. That situation was not unique and in fact had been occurring in many other midwestern and eastern states.[6]

After presenting a report at a Natural Resources Commission meeting, Fijalkowski and Rusz immediately held a press conference and in May of 2001 publicly launched what the MWC described as "field studies" to document ongoing cougar evidence.

The battle had been joined.

For the next year, the MWC regularly released enticing tidbits of information to the media heralding its field crew's findings, including evidence it had collected of tracks, cougar-killed deer, and droppings at various sites in both peninsulas.

Much of the evidence was anecdotal and had been neither peer-reviewed nor validated by the established scientific community, but it was presented to the public as proof positive that Michigan's cougars were in fact natives surviving on the fringes of the urban/wilderness interface.

Not surprisingly, the subject was quickly sensationalized, and reported sightings mushroomed, with little attempt by anyone involved to discriminate between credible and noncredible accounts. Frequently quoted by the media whenever an expert opinion was sought, the MWC and its two leaders had become synonymous with Michigan's cougar controversy, and the idea of an official "cougar cover-up" entered the popular imagination.

In 2003, the MWC entrenched itself as a legitimate, nonbiased primary source with a 54-page *Field Guide to Detecting Cougars in the Great Lakes Region*, a public seminar on cougar tracking, and a widely circulated brochure entitled "Living with Cougars in Michigan."

That's not to say that the MWC or its directors lacked professional credibility. Dennis Fijalkowski holds bachelor's and master's degrees in wildlife management and a master's degree in forestry from Michigan State University. He was a cofounder of the Michigan chapter of the Wildlife Society, the Michigan Wild Turkey Federation, five chapters of the Michigan Duck Hunters Association, and many local environmental organizations.[7]

Prior to joining the conservancy, Dr. Patrick Rusz served as an environmental consultant for 13 years and spent nine years as a faculty member and assistant to the deans at Grand Valley State University in western Michigan. He was the first president of the West Michigan Chapter of the Michigan Duck Hunters Association and has been active in many other conservation groups.[8]

But critics claimed the men were acting in a decidedly nonscientific manner, approaching their subject with a preset bias, distorting data and events, and misrepresenting their beliefs as scientific fact, thereby inciting public hysteria as a result.[9]

While many of the supposed eyewitness accounts were implausible at best—a "black panther" lurking around Southeast Michigan suburbs was a popular and stubbornly recurring urban legend—more credible evidence was also accumulating. It was getting tougher all the time to dismiss it as mere propaganda or wishful thinking.

Slowly, state and federal wildlife officials shifted their public stance to a calmly understated acknowledgment that a small number of cougars might in fact be living in Michigan's Upper Peninsula. In 2003, the National Park Service posted signs at the trailheads in Sleeping Bear Dunes National Lakeshore alerting hikers they were in potential cougar habitat.

With remarkably little fanfare, the DNR followed suit by adding cougars to its website resources on wildlife species in Michigan, along with an online tool for reporting possible sightings. Three sets of tracks were confirmed by the DNR in Delta and Marquette counties in 2007 and 2008.[10]

It's important to note that the DNR's change in its public position was—and still is—entirely different from agreeing with the MWC's base premise. The MWC asserts that the modern cougars are historic Michigan residents rather than recent arrivals via natural dispersal or human means. State wildlife biologists firmly maintain that there is still no evidence to support that claim.

There also remains a huge disparity between the two sides regarding the size and range of the population. State wildlife officials cautiously estimate that a few lone animals are mostly scattered in remote parts of the Upper Peninsula, with as yet no clear evidence of a viable breeding population.

Fijalkowski's estimates are decidedly more ambitious. In one notable 2009 interview on WJRT TV-12 with reporter Randy Conat, he was quick to peg the population at roughly 100 cougars distributed in both peninsulas of the state. While the science behind his estimate was unclear, the MWC spokesman made no bones about publicly blasting the other side, suggesting that the DNR was motivated less by science than by fear of being forced to fund a cougar habitat management plan.[11]

The truth might indeed lie somewhere between the two positions. Regardless, the whole messy affair has mostly served to weaken the DNR's credibility among the public and to reinforce the notion that state officials have for years been actively engaged in their own wildlife version of a Watergate conspiracy. By the time the Thumb wolverine showed up in 2004 and Jeff Ford rediscovered it in 2005, any statements about its origins made by the DNR were being taken with more than a grain of skeptical salt by a suspicious public.

That attitude has frustrated wildlife biologists like Arnie Karr, who feels the DNR's handling of the wolverine incident should have proven just the opposite.

"This wolverine sighting shows if we see it we'll tell people about it," he said. "When people say they know what they've seen, we know that's not always the case, but you can't tell people that. We still get criticized by certain individuals for covering up and brushing under the table cougar sightings."

Karr should know; he personally investigated dozens of cougar sightings for the DNR from 1990 until his retirement in 2010.

"I'd get two or three cougar sightings a year, and they were always wrong. They'd go to great lengths to protect the evidence, cover up the tracks with coffee cans to protect them and things like that. And they were always dog tracks," he said.

"It's an honest mistake. There have been situations when a dog might be posed in a way in poor lighting or bad conditions that I could see it happening myself."

But sometimes people continue to believe even when all evidence proves the contrary, said Karr.

"I had a call in farm country once, where a guy said he saw a large black cat with a long tail playing next to a ditch. I went right to the spot and all I saw was dog tracks. With cat, you don't see toenails. If you see toenails automatically you know it's a dog. As I was walking out of the location I heard a dog bark. I turned around and here was this skinny black dog with a long tail running back to its house a quarter mile away."

Karr followed up with the farmer to let him know the results of his investigation.

"I told him I can't tell you what you saw, but I'll tell you what I saw. And he said, okay, fair enough."

Two weeks later Karr was at a gas station in the area. The owner excitedly confronted him for more details about the "news" of a confirmed black panther sighting.

"He says so-and-so came in and said he'd seen a large black cat and the DNR had come out and confirmed it. That's how these stories get started. People hear what they want to hear."

Regardless what anyone's opinion was of the cougar controversy, the bottom line here is that the DNR's position on the wolverine's nonnative origin meant state wildlife officials were not compelled to take any further action on its behalf.

It might seem surprising to learn that the trio of amateur wolverine researchers wasn't really upset by the DNR's unwillingness to fund any further study or protection of the wolverine. The truth was that the state's official lack of interest turned out to be an unexpected gift, said Noble.

"It made it possible for us to continue doing what we were doing, to continue our research, with nobody telling us how to do it," he said. "The DNR is underfunded, and they have a lot of irons in the fire. I can understand why they would be passive about it and not eager to take on new tasks. We were eager to take it on for them."

So, with the DNR turning a benevolent blind eye, the men felt free to pursue their mission in whatever direction and by whatever means they deemed worthwhile.

Could there be another, more intriguing hypothesis for the animal's origins? Jeff Ford and his buddies were determined to find out. Now that they had documented proof that the wolverine was still alive and well and living in the Minden City bog a full year after its initial discovery, one thing was clear: even if it had started as a visitor to Michigan, it was now a permanent resident, with a home territory that should be easy to define.

A lone photograph was just the beginning. Now that they knew where to find it, there had to be a way to systematically study the creature, monitor its daily habits, and collect physical evidence.

"Once you get one picture, you want more. We stepped it up a lot then with more increased effort. We didn't really change our tactics, just a renewed vigor," said Noble. "By the time we got that first picture, the snow had almost receded. We knew we had a real short window to get any more before we'd have to pull our setup out."

On Monday, March 14, 2005, Jeff fired off a short, urgent e-mail to The Wolverine Foundation, Inc. (TWF), an Idaho-based nonprofit organization formed in 1996 to "promote interest in the wolverine's status and ecological role in the world wildlife community . . . comprised of leading wildlife scientists knowledgeable in the life history, ecology, and management of one of the least understood and most fascinating creatures on earth," according to the TWF website.

In typical Jeff Ford style, his introductory communication was simple, direct, and completely to the point.

> My name is Jeff Ford. Saturday, March 12th I went to check one of my game cameras in a remote area of the Lower Peninsula and had a pleasant surprise when a beautiful wolverine's picture was taken 6 feet from the camera, broadside. Have there been any other known cases in the lower 48 states of wolverines being caught on game cameras? What is a wolverine's life expectancy?

An enthusiastic reply from TWF came almost immediately. Administrative Manager Judy Long was already very familiar with the tale of the Thumb wolverine. Back in 2004, when the initial sighting hit the national news, the number of site visits exploded on the foundation's website, its inbox flooded with requests for more information.

> Greetings from the Wolverine Foundation, Inc. (TWF). Thank you for contacting TWF regarding your wolverine game camera photo.
>
>   We are not aware of any wolverine photos caught by game cameras, especially not in Michigan's Lower Peninsula! We are most interested in your photo for documentation of presence of the wolverine in MI. You probably heard of the documented wolverine sighting in Ubly, MI, last year about this time? It was not widely

publicized at the time, but everyone in the science community assumed that the sighting was a captive that had escaped from a Minnesota zoo about a month before, as current information indicates no viable wolverine population in MI. We have not heard of any further documented sightings out of Michigan since then.

Is it possible for you to send a scan of your photo to us via email? Please be assured that we will keep the photo only for internal documentation and will not provide it to anyone else without your written consent. Would you also provide us with details regarding exact location, weather conditions, etc. We are very interested in your documentation and hope to hear from you again regarding this issue.

Jeff wasn't sure what he believed yet—so far, there was simply not enough evidence to settle on any one particular theory—but he was intrigued at the notion that this wayward wolverine might be the escaped zoo animal from Minnesota mentioned by Judy Long. He didn't need confirmation that it was a wolverine, he told them—the DNR had already done that. What he *did* want was more information on the zoo escapee.

The TWF reply came on March 15, 2005, and further expounded on the Minnesota hypothesis. The ensuing series of e-mails between them throughout that day also provided a hint of yet another, even more intriguing possibility.

Based on your original message, my request for a photo was to eliminate the possibility of a misidentification. Normally when we receive wolverine sighting information from the northern tier states, from Minnesota to Maine, the mammal in question is a fisher. Even with photos, many people don't know the difference between the wolverine and the fisher. Since you stated that you had a photo, I felt that would be the quickest route to the next step of identification. Now I see that this situation is entirely different.

The information, as we received it, concerning the Minnesota escape is as follows. We received a telephone call from a gentleman in Zumbrota, MN, on 02/19/04. He stated that his car dealership had a wolverine on their surveillance tape from a week prior. He wanted to know the likelihood of such an event and asked if we would confirm the identification if he could get us a copy of the tape. We told him it would be abnormal for a wolverine to be in this area, especially right in town! We pointed out the possibility of an escaped captive, which we know has happened 3–4 times in the last five years or so in the NE [Northeast]—from different facilities. We didn't ever receive a follow-up from him, but then saw on the Internet news releases of the incident with a shot from the surveillance tape. It was indeed a wolverine!

On 03/17/04 I called Minnesota DNR to once again confirm their assessment of free ranging wolverine presence in their state. I spoke with Conrad Christianson, head biologist for MN DNR. Conrad stated, ". . . . there are no documented wolverine sightings in MN since approximately 1889, other than escaped captives. The recent Zumbrota sighting is most likely the Minnesota Zoo's escaped individual from early January." He also indicated that further inquiries concerning wolverine status and potential sightings in Minnesota should be directed to: John Erb.

Given the MN information, we all felt that it was a very strong possibility the Michigan sighting might be the same individual. The wolverine science community does not consider Minnesota, Wisconsin, Michigan, etc. current wolverine range, meaning supporting viable, reproducing populations. However, evidence does suggest the recent increase of wolverine numbers in Ontario, Canada, with limited sightings occurring in central and southern Ontario.

Whether this Michigan animal represents a possible disperser from Ontario, or an escaped captive individual, is unknown. It's just wonderful to hear that he is still making a living!

Up until this point, Jeff and his buddies had jealously guarded the photograph and the exact location where it had been taken, unwilling to share those facts with anyone lest the wolverine come to harm. But Jeff was completely disarmed by the friendly professionalism and openness of Judy Long and TWF. Here was an organization that could genuinely help them on their quest for answers. He wasted no time firing off another missive to Judy that day.

You've been very helpful and it's much appreciated. I will mail you a copy of the photo with the understanding it will only be used for internal documentation and confirmation and will not be released in any way. I am an outdoor writer so once the article and picture are published then I will contact you and you may release the photo at that time. The picture was taken between the night of March 6th to the night of March 11th, 2005. It was taken in the Minden City State Game Area NNW [north-northwest] of Deckerville. I have come across his/her tracks 7 times since last January, and it took me 9 months and who knows how many rolls of film before my camera finally took him. He seems to prefer an area in the game area that has a high population of Snowshoe Hares. I suspect that is a staple in his diet during winters here. In your e-mail you mentioned the wolverine in Minnesota was caught on a video in town. Had it escaped from a zoo prior to that? I'm very interested if it escaped from a zoo because I could find out if it's male or female, and its age. By the way, what is a wolverine's life expectancy in the wild? It makes sense to me that it may have escaped from a zoo because a biologist in Canada who has done extensive research on wolverines said a wolverine that was born in the wild, raised in the wild, with no contact with humans would not have run from the dogs the coyote hunters put on them near Ubly in 2004. He said they all would have been killed rather quickly. Is that one of the reasons the scientific community thinks the animal came from Minnesota? Why not from Ontario? The distance traveled is similar. Leave your address and I'll send you the photo tomorrow, in care of your name. S/he's beautiful!!!

Jeff's probing was straightforward and relentless, without guile or as much as a nod of deference to the unspoken politics and professional courtesies of the scientific community. If he was stepping on toes, he was as oblivious as the proverbial bull in a china shop. Some academics might have blown him off at this point, but his obvious sincerity and passion kept the TWF administrator engaged in the conversation. Although Jeff didn't realize it at the time, it was to be the start of several lasting friendships with the nation's top wolverine experts—a situation that would eventually put him solidly in the middle of a years-long debate within the scientific community as to the animal's origins and history. Later that day Judy Long sent this reply.

The Zumbrota, Minnesota, sighting caught on tape was the same sighting that Conrad Christianson, Minnesota DNR biologist referred to when he said, ". . . there are no documented wolverine sightings in MN since approximately 1889, other than escaped captives. The recent Zumbrota sighting is most likely the Minnesota Zoo's escaped individual from early January." We (The Wolverine Foundation) then felt there was a possibility that the Michigan wolverine sighting at Ubly may have been the same individual that escaped from the Minnesota Zoo as well. As the information we have concerning the escape is second hand, you would need to contact the Minnesota Zoo directly. I really don't even know who you would ask to speak with, or if they will confirm the escape, but the following is their contact information from the Internet:

Minnesota Zoo
13000 Zoo Boulevard
Apple Valley, MN 55124
Phone: 952-431-9200, 1-800-366-7811
24-hour information line: 952-431-9500
Fax: 952-431-9300
General Zoo Information: info@mnzoo.org

To answer your question about wolverine life expectancy: Wolverines in captivity have been documented to live up to 18 years. The life span of free ranging animals, however, is much lower, with the average life span probably not exceeding 4–6 years and the maximum life span typically not exceeding 10–12 years. Mortality resulting from starvation, predation, injuries and trapping, etc. all account for the low average life expectancy of wolverines in the wild.

Concerning the comment from a Canadian wolverine researcher; Do you have the name of the researcher. We would be interested to know who it was. We certainly do not agree with a "blanket" statement that a wild born and raised wolverine would not have run from the dogs and hunters at Ubly. There is documentation of a wild wolverine fleeing from harassment by a single coyote! Every set of circumstances will dictate a different outcome in confrontations. We all just felt that the likelihood of a wolverine in Minnesota and Michigan came more from an escaped captive. However, as I mentioned, no one can absolutely rule out the possibility that it could be an "out of area" disperser from Ontario, Canada. It could be either!

The friendly debate continued between them throughout the next day. The more Jeff thought about it, the more certain he became that this Minnesota wolverine wasn't *his* wolverine.

I did some map checking while my students were taking a test. Zumbrota is SE [southeast] of St. Paul near the 60-52 intersection. If the wolverine traveled a straight line across Lake Michigan, headed east across the Lower Peninsula to Ubly, the distance would be 500 miles. That means s/he would have had to travel 41.7 miles each day over the 12 day period. S/he would have also had to navigate many major cities during that route. I think it would be impossible that the 2 are the same wolverine. What do you think? I sent your picture out today. Let me know when you get it. By the way, I measured the prints. They are 4.75 inches long and 3.75 inches wide.

It was at this point that Judy Long first suggested the possibility of using DNA sampling to solve the mystery of the Thumb wolverine's origins. For TWF the whole question might have been nothing more than idle scientific curiosity and the use of DNA testing simply a means of answering it. But acquiring that DNA evidence would become the Holy Grail of Jeff's quest—and the unintentional fuel that would spark the fires of scientific debate in the ensuing years. Judy sent this email on March 16, 2005.

I did my own map calculations this a.m. prior to receiving your message below. I agree that it seems unlikely that the two are the same individual, not because of distance traveled, but because of the time frame limitation. One way to perhaps determine where the gulo came from would be through DNA. Zoos are documenting DNA lines and our Ontario study is also processing DNA samples to determine population relatedness, as well as many other current field research projects. Also, captive animals sometimes have identification implants. Not always though, as there are numerous "game farms," or even individuals, that hold wolverines and do not necessarily have them tagged or implanted with ID's. Of course, identification by DNA would require capture and sample collection.

I have tended to refer to "your" individual as "he" just from the original photos that appeared on the Internet of the Ubly sighting. Size estimation is very difficult, but because of head confirmation, etc., we felt it was a male. Your track measurement seems to reinforce that judgment. FYI [For your information], the following is our "average" range data for wolverine tracks. However, keep in mind that these averages are calculated from actual foot measurements of captive individuals and imprint measurements are going to vary according to the conditions of the soil, snow, etc.

Following are the range of measurements for adult male and adult female wolverines (excluding toenails):

Right hind foot:
Length: 14.5 cm (females) to 18.7 cm (males)
Width: 6 cm (females) to 8.7 cm (males)
Right Front foot:
Length: 11 cm (females) to 13.7 cm (males)
Width: 6 cm (females) to 9.1 cm (males)

Even as he was foraging online for advice and information from wolverine experts in the academic community, Jeff continued to brainstorm with buddies Jason Rosser and Steve Noble on how they could get more and better images of the wolverine. They were all avid deer hunters, but this was an entirely different kind of prey and their goal decidedly different from straight-out hunting.

What kind of bait would be most attractive? Should they stick to one spot or move the camera around? Would their frequent forays into the swamp risk driving the wolverine off? If they went public with their activities, would they be unintentionally putting the wolverine in danger from curiosity seekers or, even worse, unethical poachers who would like nothing better than a wolverine pelt to hang on the wall?

Spring was looming ever closer, putting increased pressure on them. They had one lone picture, and time was running out to get another. Steve Noble was worried about the wisdom of maintaining their presence in the woods during the upcoming spring wild turkey season.

"You don't want to have cameras lying all over the place in hunting season," said Noble. "As it was, we were always walking on eggshells about people getting an idea of what we were doing, in case anyone wanted to hurt the animal."

After a year of fruitless effort, it had become harder and harder for Noble to justify the long hours spent searching for the wolverine. But the excitement of that first photo had firmly embedded Jason Rosser as one of the team. A heavy equipment operator who lived about 12 miles from the Minden game area at the time, Jason didn't have a problem coming up with the free time to take half a day every weekend to trek out to the swamp. In fact, if there was an outing planned, he wasn't about to be left behind. The ability to spread the work out among the three of them made it easier to stay diligent with the task at hand.

A week after that first picture, the men were already furiously plotting ways to improve the odds of capturing more and better images of their prey. Jason sent this e-mail to Jeff on March 14, 2005.

I will meet you at school on Thursday when you get out. I'm looking for a fresh road-kill rabbit to take with us thinking that may be more of his appetite. I found one this morning but it was too messy. I think we need to try to pursue this nearly every weekend this summer if we can. We may have to move our cameras every two weeks if he catches on. He seems to be traveling on runways frequently taking the easier path so I think we are doing the right thing by putting our cameras on trails.

It was followed the next day by Jeff's enthusiastic reply.

I have an idea but it's risky. I have a cassette tape of a snowshoe hare in distress that I used to use up north predator hunting. I'm wondering if we set the cassette player 60 yards up wind, and hunkered down with the video camera, if we could bring him in. If we did, we would be wise to not do it anywhere near the cameras. Probably 200 yards north I know of a mature tamarack slashing where there's some visibility. Or maybe we should wait, be discreet, and try to get some more pictures first. If we were able to bring him in, he would know he was being hunted and may disappear from that area.

Jeff
March 15, 2005

Jeff,

If we made a set-up to try to get some video, I could use my video camera plus set up both of my digital cameras to 60 sec video maybe some distance away on a runway hoping he may pass by. That may be too much though because that would require repositioning both digital cameras and putting scent all over. When we eventually do this, I think we first need to find an area, get a setup made and maybe come back in a week or two after our scent is long gone and things calm back down. Do you think being in a tree would be the best place or a ground setup? I think if we were in trees we could see much further but I guess I don't know how he might react seeing a foreign object in a tree.

Jason

On March 17, their efforts paid off. Five days earlier they'd hiked into the swamp with a raw chicken Jason had purchased, hoping the novelty might entice a curious and hungry wolverine. Now, as they approached the baited camera site, they could see the chicken had vanished.

A quick trip to the Bad Axe Rite Aid film-processing desk revealed two pictures of the wolverine in broad daylight—the first shot just moments before it took the chicken and the second a few hours later after the food had been cached, a common wolverine trait. The men's whooping victory dance of high fives earned more than a few stares from the store's staff and patrons, but they hardly cared. Persistence and patience had paid off, changing the game from a single lucky shot to bona fide success as amateur field researchers. For a high school science teacher with a passion for the outdoors, that was heady stuff indeed.

Jeff was now juggling the local media behind the scenes, haggling with editors at various sportsmen's publications for first rights to publish the photographs along with his first-person account of the nine months leading up to the first successful image.

He didn't have to work hard at wooing them. Everyone was eager for a chance to publish the pictures. It was the kind of wildlife exclusive that didn't come along often in the Lower Peninsula, where the closest thing to a pristine wilderness was the large tracts of state-owned land reforested after the last century's lumber barons had stripped them clean. It's not the kind of place where many natural mysteries still lurk undiscovered.

The wolverine shows off its impressive lower teeth, which contribute to the animal's ability to crush bone. Wolverines are known to have the second-strongest bite of any animal behind the hyena.

In the end, Jeff decided to go with *Woods-N-Water News* (WNW), a family-owned publication based in Imlay City, Michigan. Begun in 1985 as a tabloid insert in the family's small local newspaper, over the years it had grown in popularity among hunters and anglers to become a full-color, 150-plus-page magazine with an audience of more than 20,000 subscribers and an estimated 80,000 other readers from 2,500 newsstands across Michigan, northern Ohio, and Indiana.

The *WNW* news editor, Tom Campbell, had purchased a few of Jeff's earlier stories on various hunting topics, including one he'd written the prior September on tracking the wolverine, and he felt some loyalty to the publication for that. This was where those precious three pictures would first appear.

On April 15, 2005, the May issue of *Woods-N-Water News* hit the stands. "The Thumb Wolverine 'Caught on Film!'" was splayed across the front cover in full color, along with photos of Jeff and his trail camera and a close-up of one of the plaster casts he and Steve had taken of a paw print.

Michigan's lone wolverine had had its official debut as the state's premier wildlife celebrity.

Early Spring, 2005

✦ ✦ ✦

By the time the *Woods-N-Water News* story broke, Jeff's old Moultrie 35mm film camera had caught three more good pictures of the wolverine sniffing out and then tearing into the meat the team had laid as bait.

In the days and weeks that followed, the story was picked up and elaborated on by newspapers and television stations throughout the state, generating as much attention as the original reports of the coyote hunters' first sighting more than a year before.

"When that wolverine was first documented by the hunters and those photos hit the Internet, our website just exploded. We got literally hundreds of e-mails too," said Judy Long, administrative manager of The Wolverine Foundation. "I guess for Michigan it was pretty exciting, being the Wolverine State and yet told for years and years they don't have any. Everybody wanted it to be part of an established population that just hadn't been discovered before now. We got so many e-mails that I had to put a posting on the home page saying this is really historic and exciting, but we're not sure what it means; we just know it's documented."

The 2005 photos and story reporting the wolverine's continued presence in the Thumb rekindled that same wave of excited interest, said Long.

But this time, Jeff Ford was no longer an anonymous Deckerville High School science teacher following the news stories from the sidelines. Now he was at the heart of it, square in the middle of the spotlight.

All three buddies had juggled television and newspaper interviews when the general media first caught wind of the story. But after the first flurry of attention, both Jason and Steve began referring reporters to Jeff as the main source of information.

"Jeff put an enormous amount of hours and money into the whole thing, far more than us with all the videos, all the running back and forth. He was pretty emotionally involved," said Rosser. "It was only natural that he pretty much took over the conversation."

Sharing or not sharing the spotlight wasn't anything the trio had really considered or discussed. But Jeff's leadership role in the project had become pretty apparent to the press and public alike.

That was the status quo by the time one of the biggest newspapers in the state came calling.

After several lengthy phone interviews, the *Detroit Free Press* sent a photographer up to Ubly to get a last-minute photo to publish with a story slated to run the next day. When Jeff heard the photographer was on his way, he immediately called Jason.

"Get your ass up here quick if you want to be in the *Free Press*," he told him. The two men posed in front of Jeff's house, with Jeff proudly holding up a plaster cast of the wolverine's paw print.

When the photographer transmitted the image to the photo desk in Detroit, the editor wasn't pleased. He wanted a solo shot of Jeff instead. Puzzled, Jeff stood his ground, arguing that Jason was his partner and as much a part of the story as he was. With the press deadline approaching, there wasn't time to dicker, and the editor relented. He had his photo, and that would have to be good enough.

The next morning, Jeff was stopped in the school hallway by Dick Walker, the principal of Deckerville's elementary classes.

"How does it feel to be famous?" Walker asked, chuckling.

A little later, another teacher, Barb Warren, stuck her head in the doorway of Jeff's classroom, waving a copy of the *Detroit Free Press*'s April 22 morning edition.

There it was on the front page above the fold: "Tracking a Wolverine through the Thumb with a Guy from Ubly"—a full-length, two-page feature article with three pictures of the wolverine, a graphic, and a map pinpointing Ubly in the Thumb. In the middle of the spread was a huge full-color image of Jason and Jeff holding up the plaster cast. Jeff, of course, was the "Guy from Ubly" in the headline and featured throughout the story. Jason was described in three short words: "another outdoor enthusiast." Steve was simply referred to as "a friend who'd been helping him."

It wouldn't be the last time the media used photos of the camo-clad trio—when asked to supply a picture of himself, Jeff typically included group shots with one or both of his partners. Occasionally the newspapers would even run them.

It didn't change a thing. Without any of them planning it that way, the "Wolverine Guy" had become a bit of a rock star in the outdoors world. And—like the lead singer in a band—Jeff's star had slowly but surely eclipsed that of his partners.

What none of them fully realized was that, as far as the press was concerned, the big story wasn't really the wolverine itself. What caught the media's attention and held it was the human interest angle: the odd tale of an ordinary "guy from Ubly" who had devoted a year of his life to studying a rare and elusive critter for no real purpose beyond his own curiosity, vowing never to quit until the day either he or the wolverine was dead. It was as though the tale of Moby Dick and Captain Ahab had turned into a quirky local wildlife love story, a writer's solid gold.

On its own, the wolverine's tale might have been a little 15-inch inside story buried somewhere in the back sections of the newspaper, where nature and science stories normally end up if they make it into print at all.

But with Jeff Ford, the quirky, obsessive "wolverine guy from Ubly," it was headline news.

Even as Jason and Steve slipped into the shadows behind the scenes, the phone kept ringing steadily in the Ford household, and Jeff was eagerly rising to the challenge. Every reporter wanted to dig up one more unique angle to explore, and Jeff was happy to oblige. With his larger-than-life personality and a natural storyteller's love for spinning a good yarn, he was an interviewer's dream.

It wasn't just reporters who came calling. Jeff had placed his phone number at the end of that first *Woods-N-Water News* article in mid-April, and people were eagerly taking him up on the open invitation to talk with the Wolverine Guy in person.

Almost as soon as the issue hit the stands on April 15, the calls began to roll in. After the first week, the Fords gave up and let the answering machine take over, adding wolverine queries to order requests for Amy's home makeup sales business. It might have been the first time Mary Kay Cosmetics shared billing with a wild wolverine: "Hello, this is Amy. We are not home right now but anyone calling about the wolverine or Mary Kay, please leave a phone number and we'll get back to you."

It turned out that all the media attention was bringing an interesting variety of people to Jeff Ford's door, each with his or her own agenda— from oddballs seeking confirmation for their own outlandish "eyewitness" accounts of improbable wildlife encounters to sportsmen, researchers, and wildlife conservationists.

Within the first three weeks after the article ran, they received more than 50 claims of sightings, 30 requests for interviews, and even a few polite inquiries from the scientific community.

The purported wolverine sightings were by far the most entertaining and ran the gamut from simply odd to utterly outrageous. Like jackrabbits in heat, it seemed that one confirmed wolverine case was all it took to spawn dozens of copycats across the state.

One Grand Rapids caller claimed a wolverine had shot across the road in front of him; when he got out of the car to confront it, he said, he'd narrowly escaped with his life after it lunged at him, growling and hissing. Another claimed a wolverine was living under her front deck: would Jeff please come over and remove it?

A year ago, anyone would have said the chances of finding a wild wolverine anywhere in Michigan were pretty much nonexistent, nothing more than a naturalist's wishful thinking. Now, like a rampant case of the emperor's new clothes, it seemed everyone was seeing these rare, elusive creatures wherever they looked.

Either the Thumb wolverine was the fastest-moving land mammal on the planet or an entire underground population had been lurking in the shadows for 100 years, awaiting this one moment in time to launch a massive coordinated attack on suburbia.

Still, Jeff dutifully listened to them all, politely answering their questions and thanking them for sharing their stories.

Other people were requesting photos, and it didn't take long for Jeff to realize the potential for a small moneymaking enterprise on the side, churning out autographed 8 × 10 photos.

Local schools and sportsmen's clubs were calling too: would Jeff be willing to give a talk to their group and share some images of the wolverine?

The wolverine explores the research site.

It was all getting to be too much. At Riley's third birthday party in April, Jeff confessed to his sister Teri that he was feeling overwhelmed. She wasn't surprised: even up in Saginaw, where she lived and worked as a Web designer, several people had confronted her with questions about her brother's adventures with the Thumb wolverine.

It's time for a website, she told him—and she was just the girl to do it. It would be the ideal way to keep people informed without infringing on his personal life, giving him more time to do what he needed to do: study the wolverine.

Jeff immediately began funneling information to her, and a couple of weeks later www.wolverineguy.com (now www.wolverineguy.org) was up and running. Now he had a "Latest News" section in which to post new information, another place where he could share his favorite photos and links to his latest articles, and even an online shop where people could order unique products such as plaster casts of paw prints and full-color photographs. Now whenever he did an interview or wrote a story he'd refer readers to the website instead of offering up his personal contact information. It worked just as Teri had promised: now when the usual flurry of queries came in after a new story had run, they were herded straight to the website instead of landing in a heap on his home phone.

Within a few months, the site had garnered over 40,000 hits. In its own small way, the Thumb wolverine was an Internet hit. Not bad for a furry weasel living alone in a swamp.

This not only served to keep the Thumb wolverine in the public consciousness but also firmly cemented in people's minds Jeff Ford's central role in the animal's life story.

The wolverine was still a wild animal, free of anyone's claim of ownership, but in a very real sense Jeff Ford had gained an unspoken proprietary role in its continued existence and media coverage. By this time, most journalists wouldn't even consider writing about the wolverine without consulting Jeff Ford first.

Somehow, without any official acknowledgment or permission, the Thumb wolverine had become "his" wolverine. There was no denying that Jeff loved the attention and notoriety of being Michigan's Wolverine Guy. But—like a prospector guarding his claim—he also became increasingly wary of anyone who showed *too* much interest. He was painfully aware that the spotlight he'd placed on the wolverine could also put it at greater risk.

The ones who worried him most were the local nature enthusiasts or fellow outdoorsmen who wanted a piece of the action for themselves—and they weren't always willing to take a polite no for an answer. More than a few resented the proprietary stance taken by Jeff and his buddies. It wasn't on private land owned by any of them. They weren't even the original discoverers. They had no official authority from the state, no scientific or academic status. What right did they have to treat the wolverine like their exclusive personal property?

One caller, a man named John from Harbor Beach, kept him on the telephone with an endless barrage of questions about the wolverine. At first, Jeff was happy to share stories of his forays into the swamp. But the conversation kept circling back to the same request: could the caller come along the next time Jeff went out to visit the wolverine? Again and again, Jeff politely but firmly refused, explaining the importance of keeping the wolverine safe from unnecessary human intrusion.

When he realized Jeff wasn't going to give in, the man's disposition changed from enthusiastic to angry and threatening. The last thing he said before slamming down the phone was that he was going to find that wolverine on his own, and nobody was going to stop him.

Unwelcome encounters like these left Jeff feeling bruised and wary, but they only strengthened his resolve. This was just one guy. How many others like him were out there just itching for a chance to exploit the wolverine for their own purposes or even to kill it as a trophy?

It wasn't an easy dance to keep up without tripping over his own boots. On one hand, Jeff was aggressively promoting and publicizing everything he did concerning the wolverine; on the other, he was jealously guarding those activities from outsiders' prying eyes. He knew one careless mistake could allow someone to follow the men through the swamp to the remote research site at the heart of what they'd come to call its "sanctuary." Blowing wide open the secret of the wolverine's home territory would expose it to things far worse than trail cameras.

It wasn't just a case of paranoia with no grounds in reality. Like the Harbor Beach caller, there were more than a few instances when Jeff knew without question that was exactly what was going on.

Shortly after the first news story broke, he was driving home from school when he noticed the same car had been tailing him for quite some time. Suspicious, he began making some evasive turns, stretching the 20-minute drive out longer and longer. The car stayed with him. Finally, he slowed to 20 mph, hoping to make it obvious to his pursuer that he was aware of him and the game was up. It was only then that the frustrated driver hit the gas and sped around him, vanishing from sight.

On another occasion, in early May, Jeff had just returned home from the swamp and was sitting in the living room playing with Riley when he noticed a red truck slowing as it cruised past the house. A few minutes later there it was again. The Ford house is in a very rural area about a quarter mile from the swamp on a little-used stretch of road— it was unusual enough to see anyone drive past, let alone the same vehicle, again and again. He counted as the red truck made three more passes before finally continuing on its way down the road.

He didn't like the feeling of being stalked by strangers, but he wasn't really worried that anyone meant him or his family harm. What concerned him was what they might be learning about his comings and goings in the swamp.

That spring there was still snow on the ground in March and April.

Anyone who knew his vehicle and found it parked along the perimeter of the swamp could wait for him to leave and follow his tracks in the snow directly back to the research site.

Keenly aware of that risk now, Jeff started taking special pains to make sure no one followed him on his forays to the swamp. Once he was sure no other vehicles were around, he'd head out to the end of Palms Road, then park and hike in through the swamp with his load of fresh bait.

He also began covering his trail. Using a small handsaw, he'd cut off a stout tamarack limb with a few good branches on the end. Walking backwards, he'd brush the branch over the snow to obscure his tracks. Once he was far enough in, he'd begin dogging back and forth through the heavy brush, scrambling through and around the thickest, most impenetrable scrub he could find. If anyone did try to follow him, he reasoned, they'd be in for one long, miserable day.

The easy trips were on days when he headed in on the leading edge of a snowstorm and could feel confident that the wind and snow would cover his tracks for him.

Later in the spring, after the snow had melted, he still had to deal with obscuring his path through the mud, which could be even tougher. A wrong step could sink his foot up over the top of his knee-high boots—not an easy mark to wipe out.

It all added long hours to the regular treks, and made each hike an exhausting labor. But it was worth every hour and bead of sweat if it kept someone from trapping the wolverine for a trophy. He wasn't naive enough to think his efforts were a perfect form of protection, but at least they eased the nagging guilt of knowing his own actions might be the biggest threat the wolverine faced.

Amy Ford just took it all in stride. Once, when Jeff apologized for all the long hours he was spending away from her and the kids, she simply responded, "Better to have you chasing something on four legs than two." Jeff laughed, smart enough to leave it at that.

"Being that I was teaching, hunting, fishing, and spending a lot of time in the swamp pursuing the wolverine, many wives would have probably called it quits with me," he said. "Maybe part of it was the fact she knew I was going to do it anyway, regardless, so she might as well just accept it, deal with it, and forge on. Anyway, I respect her for allowing me to pursue all my outdoor passions."

Jeff Ford maintains the research site, changing memory sticks, replacing batteries, and baiting.

Michigan DNR officials, meanwhile, remained interested but uninvolved.

"At that time there were no baiting bans like there are now, and, of course, we did advise him to stop using chicken. I told him, 'Let's not accustom this animal to eating domestic livestock,' and he agreed to that. After all, we didn't want the thing raiding the KFC in Bad Axe someday," joked Karr. "So as far as we knew, he was just taking game meat products out there, and we really didn't see any harm in that."

The unspoken understanding also gave the DNR a free extra pair of eyes with which to monitor the situation.

"It was helpful to us in knowing it was still there," said Karr. "In talking with the public, people always asked us how the wolverine was. We could give people an answer that way."

Aaron Schenk and his buddies, who'd originally found the wolverine, weren't so tolerant at first of Ford's growing notoriety.

"In the beginning when we heard about Jeff doing his cameras and pictures, I was a little resentful of what he was doing and how he was going about it," admitted Schenk. "We didn't think what he was doing was legal, with feeding it and taking pictures all the time, being on it like that. It didn't seem like it was in its natural state anymore."

Slowly, over time, that annoyance turned to a grudging respect and eventually admiration.

"All of us have bought every magazine or newspaper whenever Jeff had an article on what he was doing. We were all interested in keeping up with it and appreciated the effort and work he put into keeping on it," said Schenk. "This will probably never happen again. Thanks to him, it's all on record."

Late Spring, 2005

✦ ✦ ✦

By this time, Jeff had acquired over 30 individual shots of the wolverine, using the same baited camera technique, and he was busily farming them out to whatever media outlets were interested in publishing them.

Jason hadn't been so lucky with his newer digital Cuddeback model. True, he'd managed to capture the image of a bald eagle gliding across a clear blue sky—but, unexpected and magnificent as that might seem, it was hardly the same thing as capturing a shot of Michigan's only known wild wolverine.

Once on the brink of extinction across the entire lower 48 states due to such factors as habitat loss and pesticides containing PCB and DDT, the bald eagle had seen a slow but steady resurgence ever since the enactment of the Endangered Species Act in 1973. Now they nest mostly in the Upper Peninsula and the northern half of the Lower Peninsula, with notable exceptions including an established nesting population in the Shiawassee National Wildlife Refuge near Saginaw. With their wide foraging range, bald eagles can now be seen during the winter in almost every county of the state.[1]

So getting a picture of a bald eagle while scouting for a wolverine? Close but no cigar.

Jason suspected the severe cold was causing the camera's batteries to slow down or fail. He built an insulated box that could be screwed to a tree, with a hinged door and padlock. It was only a matter of time now, he felt, before the wolverine crossed his camera's path as well.

Capturing good images of animals in the wild is always a challenge for researchers. Jeff and Jason had been learning the hard way that it is often necessary to devise and change strategies on the fly as the situation demands, almost like a game of wits between researcher and subject.

The main challenge was making sure the equipment was in the right spot at the right time. As often as they were successful, it was just as likely that the partial images they captured only revealed tantalizing hints of what they had missed.

But all that was about to change.

On April 3, Jason called with exciting news. He'd just come back from the research site where he'd collected some amazing images on the memory card of his trail camera. At 10:01 a.m. on April 1, the lens had recorded a picture of two red-tailed hawks perched on top of the venison pile. One minute later, at 10:02, the camera had captured the wolverine in midair, sprinting toward the venison pile.

To Jeff and Jason, the scenario was obvious: they'd caught the wolverine in action, chasing a pair of feathered competitors off its food cache.

A week later the pair met early and started their hike in, more excited than usual. Because of the hawk incident, they decided to approach the site very slowly and quietly, as if they were deer hunting, in hopes of actually seeing her. Jeff led the way with Jason close behind. Crossing the last 150 yards, they'd take three quiet steps, pause for about 10 seconds, then take another three steps. A southwest wind was blowing into their faces all the way in, a good sign. They knew the wolverine wouldn't be able to catch their scent.

Once at the site, they quietly started changing camera batteries and swapping out memory cards and film. Because he was still working with the clumsy old 35mm model, it took Jeff longer to finish, so Jason was standing behind him, waiting. Just as Jeff was snapping the film into place, he heard him whisper, "Jeff, I hear footsteps."

He was pointing toward the southwest, but they could only see about 10 yards into the thick underbrush, with a few scattered holes where the visibility stretched no more than twice that distance.

Both men remained frozen in place, listening. The sound of steps came again, maybe 50 yards off.

"That's a deer, I think," whispered Jeff.

Then suddenly, the slow, tentative steps exploded into the sound of cracking and crashing brush. Whatever it was, it was charging straight at them!

Seconds later the wolverine burst into the clearing, mere yards from the men, obviously hell-bent on chasing off whatever was threatening its food cache. But as soon as it saw the two humans, it stopped in mid-

The wolverine's muscular buttocks and tan stripe make it clear why one of the animal's nicknames is "skunk bear."

charge, bolting to the right nearly in full reverse and vanishing back to the southeast.

It all happened so quickly they didn't have time to react. Almost before their stunned brains had processed the event, it was over.

Jeff felt a chill spread across his skin. He glanced over at Jason, who was still frozen in place, eyes huge and mouth open.

Jeff grabbed his gear, still shaken by what had just occurred.

"Let's get the hell out of here," he said. "We'll talk later."

They didn't stop moving until the undergrowth began to thin out around them, the research site left far behind. Then the words came out in a rush, both men pacing in circles, swearing, and laughing as they excitedly compared notes on what they'd just witnessed. Minutes later Jeff was on his cell phone relating the tale to Steve Noble. But it wasn't until December of the following year that anyone else heard the story beyond that close-knit circle of three. Wolverines already had enough of a public relations problem, they reasoned, the species' image distorted into that of a fiercely ill-tempered, vicious predator with few redeeming qualities. Like most Michiganders, even Jeff himself had unwittingly bought into much of that mythology before firsthand observation and careful study had taught him otherwise.

While powerfully built and well suited to rugged winter climates, wolverines are in truth mostly scavengers rather than hunters. Even though the popular literature is filled with tales of wolverines fighting off competitors, they are relatively small compared to large predators such as bears, wolves, and mountain lions and are known to have been killed in such an encounter.[2]

In fact, rather than posing a danger to people, humans are among the main causes of wolverine mortality due to trapping, hunting, and vehicular collisions.

Despite all that, Jeff worried that no matter how carefully he framed it the story would still be spun in a way that perpetuated the myth. From there, it would only be a short step to a public outcry demanding the destruction of a "dangerous" animal on the loose. For the time being, at least, the secret would stay safe in the swamp.

On May 8, 2005, Jason sent three new pictures to Jeff that he'd collected the previous day. Still, it seemed like a meager return for all their efforts, as he noted in this e-mail.

He only came in the two days after we put bait out and never came back all week. We need to cover every possible approach to the bait so we won't miss anything. In order to do that we don't have enough woodland cameras so I think we are going to have to leave our cameras, otherwise we can't cover every approach and I don't want to miss any pictures because our cameras weren't there. It would be different if we had enough woodland cameras but since we don't, pulling our cameras would leave too many gaps for him to slip through. Maybe this week we should move the entire set-up to another area close by that would change everything so he wouldn't know which way to avoid any cameras?

By now they were getting in pretty deep for something that had started on a whim. It had been more than a year since they'd begun this odyssey. While Steve Noble was beginning to pull back from active participation due to heavy work commitments as a school administrator and coach, Jeff and Jason were pushing ahead even harder, taking valuable time away from their families every weekend. And it wasn't just time they were spending. It was money too.

New equipment was costly, but it was painfully evident to the men that Jeff's old trail camera was hopelessly outdated, while Jason's digital seemed too temperamental for the cold. If they were going to make any real progress, they knew they would have to make some investments in better equipment.

They tried to swing a deal with some of the companies producing trail cameras, hoping for some sort of corporate sponsorship that could fund what was gradually evolving into something more than an idle hobby. While they were more than happy to offer technical support for the purchase of their products, a few polite responses made it obvious that there would be no relief from that direction. Jeff was making a few bucks selling wolverine photos and articles, but it was hardly enough to compensate for the time and money they were pouring into the effort. Was it really worth it?

At this point they didn't stop to question it. If they decided they needed something, they simply bought it. While the brunt of the expense was borne by Jeff and Jason, Steve Noble also contributed.

"Without a doubt the equipment was a big thing, maintaining the research sites with bait, the travel expenses. It cost a lot more than a couple hundred a year," said Noble. "Was it hard to justify the expense? Not for a couple guys who blow money left and right on deer hunting, fly-fishing, and a lot of other outdoor stuff. We've been justifying it for years for other hobbies. This one was just as easy."

Here's a short, itemized list of some of the main equipment expenses in 2005.

Moultrie 35mm game camera: $150 (Jeff)
Cuddeback digital game camera #1: $200 (Jason)
Cuddeback digital game camera #2: $200 (Jason)
Cuddeback digital game camera #3: $180 (Jeff)

Leaf River digital game camera: $200 (Jeff)
Wildlife Eye video system #1: $850 (Jason)

That wasn't counting all the ongoing costs involved in bait, batteries, camera repairs, photo processing, digital compact discs, and travel expenses.

Meanwhile, Judy Long of The Wolverine Foundation had made good on her promise to forward Jeff's request for an expert opinion to her husband, Clinton D. Long, the cofounding director of the foundation. His reply arrived in Jeff's inbox on March 29, 2005.

Thanks for sharing your observations and interest in what appears to be an anomaly in your area. Although I do not have enough data to provide a definitive treatment to your questions, I would offer the following thoughts.

• *The possible origin of this wolverine:* Obviously, the closest documented, reproducing, free ranging wolverine population occurs in Ontario, Canada. It is conceivable that this individual represents a disperser from that population. However, I am not cognizant of the number of facilities, whether zoological gardens or private game farms, located in your state or the surrounding states, that hold captive wolverines. I also am not aware of the reporting criteria, whether mandatory or voluntary, for escaped individuals. Therefore, with the information I have, whether this animal is an escaped captive or an Ontario transient cannot be determined. Possibly a more definitive location of origin could be concluded with a genetics assessment.

• *6,000 acre home range:* This is significantly smaller than the smallest* home ranges reported for both adult female and adult male wolverines (*females—approx. 25,600 acres, *males—102,400 acres). Factors dictating home range size include available biomass and presence of potential breeding partners. This is especially recognized as a factor of home range development of adult males. The possibility of this area, or even the 97,000 acre state forest, being adequate for sustaining a viable wolverine population is extremely unlikely.

• *Calculation ratios to determine male or female:* This is an interesting idea that possibly could prove somewhat useful with extensive testing. However, the field standards for collecting morphological measurement data would make it difficult to ascertain sex. For example, the SOP for length measurement is measuring ventrally from tip of rhinarium to base of tail with the animal laid flat on its back. Also, some large adult females exhibit common measurements with small adult males. When pelage differences are factored in, it would be difficult to defend a sex designation based on a photo. *NOTE:* The track measurements you provided, especially width, suggests this individual is a male.

　• *Weight estimation:* Again, it would be difficult to estimate weight from a photograph. However, your estimate of 30–35 lbs. for an adult male would fall within the normal range. A 50 lb. free ranging adult male wolverine would certainly be a rarity.

　I'm sorry that I couldn't be more definitive in my responses. However, as you are aware, science assessments are based on defensible data. Thanks for your inquiries and we hope your article prompts interest in the wolverine.

Long's assessment, while informative and intriguing, was understandably inconclusive. There was simply not enough solid scientific data for anyone to do more than make a series of educated guesses. It only whetted Jeff's appetite for real answers and further convinced him that he needed to do whatever it took to get the necessary data.

Around this time, Jason read an online article written by a man named Jeff Copeland, a highly regarded wolverine expert at the Rocky Mountain Research Station in Montana, home to one of the few viable wolverine populations in the contiguous United States. Although Jason didn't know it at the time, Copeland was also a peer of Clinton Long.

"One of reasons the foundation started was this group of biologists who wanted to see accurate scientific information about wolverines put forth, and then be able to escalate on that with current ongoing research," said Judy Long. "Previously on the Internet you could only find repeated popular info from the fifties and sixties and much of it was not scientifically accurate. Even today you can go on some websites that have info on wolverines and they state all kinds of misinformation."

Obviously, Copeland was at the forefront of wolverine research. Why not reach out to him for advice? Jason took it upon himself to contact Copeland, seeking advice on how they could take their project to the next level. On April 5, 2005, Jason sent this e-mail to Copeland.

If I have done my research right you were in a segment on National Geographic some time ago on wolverines. I have recently obtained it from a friend of mine. In the past year there were reports of a Wolverine in the Thumb of Michigan. Fellow outdoorsman Jeff Ford did an article in Sept.04 in Woods and Water Magazine on the findings of the wolverine tracks he found in a remote area of the thumb. Recently we returned to the location of the tracks to try to pick up his trail again, after a few hours we had success. Over the following weeks we placed four trail cameras in the woods and some chickens bought from the store. After three weeks now we have 30 pictures of him. Jeff has been in contact with the Wolverine Foundation with our findings. We realize how rare this is so we have kept the location to ourselves. Our goal is to keep track of this animal for as long as he remains in the large swampy area. We are cutting back on the feeding so he doesn't get dependant on us. In one week our story hits the front page of Woods and Water Magazine so I am sure we are going to get many phone calls.

Just before the story breaks we are going to contact the DNR with limited information on the location. With limited resources we don't have the means of tracking this animal electronically although we are checking to see what we can do. This particular wolverine is believed to be the only wolverine in Michigan in well over 100 years. Do you have any advice on what we should or shouldn't be doing during our effort to keep track of the wolverine?

Copeland's initial reply, on April 6, 2005, was courteous but understandably cautious. After all, the two men were complete strangers to him and obviously lacked the requisite professional or academic credentials to be peers.

I'm a little hesitant to get too involved in this without some consultation with Michigan DNR. I would assume they are interested in the status of this wolverine so I would encourage you to keep them aware of what you have been doing. From my perspective I would be most interested in trying to understand where this individual came from. Wolverine are certainly capable of moving long distance so there is some possibility that this individual came into Michigan from somewhere in Canada, but it seems more likely that the animal is an escaped captive. One could possibly attain some insight into the animal's origin through DNA analysis.

Regardless, the animal is ultimately the responsibility of Michigan DNR so I would urge you to contact them. If we could be of any assistance, I would be happy to visit with the DNR folks and help in any way I can.

In his near-obsessive research, Jeff had come across Copeland's name more than once among the small pantheon of scientists who'd devoted their careers to the study of the North American subspecies of *Gulo gulo*, genus *Gulo*, family Mustelidae, order Carnivora, class Mammalia.

"It didn't take me long to realize that these people committed their entire lives to the study of *Gulo gulo*, and that is no easy task," said Ford. "They are probably the most difficult species to study, and yet year in, year out, you'll find these folks in the deep woods somewhere studying these creatures."

In fact, the more he read, the more that vast body of academic, peer-reviewed work dwarfed his own amateur research efforts. It humbled him, but at the same time it fueled his determination to somehow add to that body of knowledge, even if only as a minor footnote on one solitary, anomalous wolverine.

The two who impressed him most were Copeland and Audrey Magoun, a wolverine specialist with the Alaska Department of Fish and Game.

He'd first glimpsed Copeland's Idaho research as part of a documentary he'd seen on television and videotaped for his high school science classes. He was enthralled with the images of Copeland pushing deep into the woods in the summer to make his preparations, chopping down trees to construct a live trap out of logs large enough to hold a wolverine, which could then be collared with a GPS tracking unit and released after blood and tissue samples were recorded.

"He made a bunch of these traps, and I started getting an idea of the kind of commitment he had made to provide the research needed to determine if wolverines should be protected on the endangered species list," said Ford. "The Feds won't protect anything there is not enough data on, and people like Copeland are data-producing machines."

But when he shared the video with Steve Noble, neither man was prepared for Steve's reaction.

Back on November 13, 2004, Steve had been in the swamp deer hunting. On that particular morning, he'd just finished his hour-long walk to his tree stand. It was still an hour until dawn when he arrived and began his ritual of changing out of his sweat-drenched clothes and into his hunting gear, then climbing up into the stand 18 feet off the ground in a tamarack tree.

He made a few fawn bleats on his A-way Bowgrunter Plus hunting call to disguise the noise of his preparations, just in case a wary white-tail was nearby.

Just as he laid out his last sequence of calls, he heard a return noise off to the south, at least 100 yards away in the dense swamp. He didn't recognize the noise, but it sounded like a sick animal. Just the weekend before, he'd put an arrow into an eight-point buck but had been unable to recover the wounded animal. Was it the buck coming back to his core area to bed up for the day after an agonizing night struggling to feed and survive?

Steve could do nothing but sit quietly and listen in the dark. The noise kept getting closer. It sounded like a series of intense exhalations, like an asthmatic or chronic smoker running a marathon. The sound got louder as the unseen animal closed the distance. The noise had changed to a series of huffing, growling sounds now. At this point he knew that whatever it was, it wasn't a white-tailed deer—and it was right in front of him, barely 10 yards away, the brush breaking as it approached.

There was nothing he could do but sit and wait, praying for daylight. Five minutes passed, then 10. Nothing. Whatever it was, it was gone.

Gone, that is, until now, as he sat watching the video of Copeland's research. In one segment, the crew approached a box trap with a live wolverine inside, huffing and growling to ward off the humans' advance. The sound was suddenly, startlingly familiar.

Without even realizing it until that moment, Steve Noble had had his own encounter with the Thumb wolverine.

What really fired Jeff's imagination, though, was the story of one young male Copeland had livetrapped and collared. According to the transmitted data, the lone wolverine had covered 258 miles in 19 days.

And this was while climbing up and over mountains that were over 12,000 feet high! Jeff was awestruck at the implications. The GPS-plotted mileage didn't take into consideration elevation changes, so he figured the wolverine could have covered several times that distance on flat land such as the Thumb.

Obviously, long-distance journeys were no problem for a wolverine. It seemed entirely feasible to Jeff that his own wolverine might very well have crossed the ice into Michigan from northern Ontario, perhaps in search of a territory to call its own. After all, that's how most species dealt with growing populations, shrinking habitat, or dwindling food supplies, wasn't it?

Take, for example, Michigan's black bear population. According to the DNR, approximately 15,000 to 19,000 black bears can be found roaming the hardwood and conifer forests of northern Michigan, with about 90 percent of them in the Upper Peninsula and most of the rest in the northern Lower Peninsula. However, as more people move to northern Michigan and undeveloped bear habitat declines, bears are being seen far more frequently in the southern half of the state, even in urban areas, as youngsters are pressured to spread out ever farther and seek out territory for their own home range.[3]

Could something similar be happening with wolverines in northern Ontario? Through his studies of the available literature online and elsewhere, Jeff had learned that northern Ontario's wolverine population had been expanding for the last several years, possibly as a result of the Canadian province's Lands for Life program. Formally known as the Living Legacy Land Use Strategy, the Ontario Ministry of Natural Resources had begun in 1999 to strategically manage 39 million hectares of Crown lands and waters covering 45 percent of the province in ways that would minimize the impact of invasive land uses, such as mineral and timber extraction, in a conscious effort to protect natural resources and wildlife habitat.[4]

Too, wolverine populations in many areas of Canada were benefiting from the cessation of wolf poisoning and wolf control, restrictions on wolverine trapping and hunting, and rising numbers of moose and caribou, whose carcasses serve as a mainstay of the wolverine's typical winter diet. A 2003 report from the Committee on the Status of Endangered Wildlife in Canada (COSEWIC) noted population increases in northwestern Ontario and Manitoba, where caribou have increased.[5]

Jeff had learned from The Wolverine Foundation that there had been a number of unconfirmed reports of wolverines or their tracks in the Thunder Bay area. The only confirmed records are a male wolverine trapped about 56 miles (90 km) west of Thunder Bay in November, 1996, and a male trapped in January, 2004, about 84 miles (135 km) north of Thunder Bay.

Jeff had also learned that male wolverines will not tolerate other males in their home range, constantly scent-marking their territory to drive out younger or less dominant male competitors.

If Copeland's radio-collared male could travel 257.5 miles in 19 days, surely the Thumb wolverine could have covered the 470 miles from Thunder Bay to Huron County, Michigan, given the time and motivation. At the same rate of travel, Jeff reasoned, the Thumb wolverine could have been driven out of northwestern Ontario by a more dominant male sometime in December of 2003 or early January 2004, traveled around the eastern shore of Lake Superior, and then headed south to cross the ice of Lake Huron in time to arrive in a woodlot near Aaron Schenk's home on February 24, 2004.

In fact, who was to say the wolverine hadn't arrived a year earlier and simply remained undetected until the Schenk brothers found the track in 2004? Satellite photographs taken on March 9, 2003, by NASA's MODIS Rapid Response Team showed ice almost completely covering lakes Superior, Huron, and Erie as well as the northernmost fifth of Lake Michigan—a rare winter ice-over that would have provided multiple potential routes.[6]

There were far simpler, more elegant solutions too. Although Jeff couldn't know it at the time, researchers in 2006 would spot wolverine tracks in Wakami Provincial Park about 250 miles straight north of the Thumb, far south of "known" wolverine range. Was it possible wolverines had also roamed that area two years earlier? And that one lone wolverine had wandered even farther south, across Lake Huron to the Thumb? Satellite photos taken on February 19, 2004, showed ice covering wide swaths of shoreline and vast portions of both the north and south ends of Huron.[7]

There was also the still unresolved theory that the Thumb wolverine might be the Minnesota zoo escapee. On February 12, 2004, that escaped wolverine had been caught on a surveillance camera in the town of Zumbrota, 40 miles southeast of Minneapolis. If it was the same wolverine spotted in Michigan's Thumb on February 24, the animal would have made a 500-mile trip around Lake Michigan and through many heavily urbanized areas in 12 days' time—or migrated up through Northern Minnesota and the UP before crossing to lower Michigan at the frozen Straits of Mackinac. Both routes seemed highly unlikely to Jeff in the short time span. Still, the Minnesota and Michigan wolverine events were coincidental enough to bear serious consideration.

It seemed to Jeff that the only real way to resolve the matter was through genetic analysis, which could link the wolverine to a specific population. He was eager to ask Copeland some questions of his own.

Undaunted by Copeland's polite but firm refusal to Jason, explaining his unwillingness to get involved except through official channels via the Michigan DNR, Jeff fired off an e-mail of his own. He hoped to convince the scientist that their efforts were indeed legitimate. Copeland's mention of DNA analysis was exactly what he was looking for. On April 11, 2005, he jumped into the conversation with the following e-mail to Copeland.

I talked to Arnie Karr on the phone this morning and I am sending him a picture of the wolverine for documentation. I've been looking at the procedures in research studies for barbed wire hair traps for DNA analysis. The procedures are not listed in great detail, and I am a bit concerned about setting it up wrong and harming him. Is this something that is feasible for someone with no experience or should I discard the entire idea.

We've been finding large scat in the area, and quite truthfully are not sure if it is wolverine or coyote. The one I found yesterday consisted of 2 pieces, each 4 to 4.5 inches long, round, with a diameter of 1.2 to 1.5 inches. Can DNA analysis accurately be determined through scat, or is hair needed? Could you offer any insight into a wolverine's scat as far as dimensions, shape, color, etc.? Also, since I've spent the last year casting tracks of this wolverine and trying to get his picture I have become intrigued almost to a level of obsession with the animal, and any research studies or information related to the wolverine I read. Are there any data or studies that you recommend to educate me further? I've read studies where you were referenced and watched the special on Discovery when you were log box trapping. Great stuff! Are there any videos out there that I could purchase on the wolverine?

Copeland replied the same day. Despite his continued caution, the scientist's interest had obviously been piqued. After all, what researcher could resist someone else's obvious passion for his subject?

> I am assuming Arnie Karr must be a DNR person? If you haven't looked at the Wolverine Foundation website yet, it is a pretty good source for wolverine info. As far as hair grabbing devices, barbed wire generally works okay, and I wouldn't worry about hurting the animal. I doubt he would let that happen. We can get DNA from scat but it is very difficult to distinguish wolverine scat from that of coyote or fox, other than by smell. Our best bet would be to collect some hair. If he has been coming into a bait site on a regular basis, I would imagine there would be some hair at the site. The other option is to put bait on top of a post wrapped loosely in barbed wire. I have some other devices that are effective as well, and I would be happy to send you some if the DNR folks give us approval. I could also send some containers to put the hair in as well.

How to explain that the Michigan DNR really had no interest in supporting or preventing the little "research project" that he and his buddies had launched? In Jeff Ford's mind, it seemed like a minor point, hardly worth dwelling on. On April 12, 2005, Ford wrote back to Copeland, brushing aside the issue like an annoying fly buzzing around his head.

> Arnie Karr is the DNR Biologist that confirmed on site that the animal is a wolverine last February 24th after the coyote hunters dogs ran and treed him. I am going to try the pole with loose barbed wire after we collect pictures for a few more weeks. Would a small zip lock be OK for collecting hair samples or do I need to get a special container from you?
> Would I send the sample to you or directly to the lab for analysis? I'll also be sure to carefully check the feeding site for hair. We moved the site last weekend to a bit more secure and remote area because a Michigan magazine is coming out with his pictures plas-

tered all over the front cover, and we fear there will be people looking to find him and our cameras. It's a 1.4 mile walk to the spot, but
we just wanted to make sure. The area he is living in is a bog area
and extremely thick. He seems to like it there because he's been in
and around the same area for over a year now. I e-mailed Clint and
Judy Long the dimensions of his track, and they think he's a male
because his track width averages 3.7 to 3.8 inches. I only measured
those where the impression was inch or less to avoid error, and most
were taken in the firm mud last June where accuracy is good. Could
it be expected for him to go looking for a mate during the upcoming
breeding season? It only took him a few days to start hitting the new
site. We took 6 additional pictures at the new site. Thanks again for
taking time from your busy schedule to help us.

But Copeland wasn't about to be distracted by Jeff's casual disregard
for the proper rules of engagement between state and federal wildlife
officials. Copeland's reply to Ford on April 12, 2005, was insistent and
unequivocal, leaving no room for misunderstanding: without the
DNR's input, Jeff Ford would get nothing further from the Rocky
Mountain Research Station.

But even so, Copeland couldn't resist offering up a few tips for collecting valid samples.

Any hair samples need to be stored in desiccant. Moisture acceler-
ates the degradation process so you definitely do not want to store
hair in plastic bags. We would be happy to look at the hair samples
but, as I mentioned earlier, not without the involvement and bless-
ing of Michigan DNR. If you could provide me with some contact
information for Mr. Karr, I would be happy to make the contact. I
don't mean to be obsessive about this but this animal is ultimately
the responsibility of Michigan DNR, and as a Forest Service em-
ployee I am not going to get involved in the collection and analysis
of samples without coordination with the state. I am happy to help
however I can, but Michigan DNR has got to be aware of this before
I can proceed.

On his end, Jeff Ford barely noticed the impasse. At the same time
that he was writing Copeland, he was also reaching out via The

Wolverine Foundation to Audrey Magoun, the researcher whose Alaskan field studies had so intrigued him.

From what he'd been able to find online, in scientific journals, and in general news articles, Magoun had been studying wolverines for nearly three decades and was widely known and respected in her field. It wasn't uncommon for other wolverine researchers to reference her past work in the notes in their own articles published in scientific journals.

After years of working in the field, Magoun wondered whether the study of captive-born wolverine kits reared in a natural environment could provide new insights into the development and ecology of wild wolverines.

Her goal wasn't simply to satisfy idle academic curiosity. Increasingly, wildlife biologists had come to realize that the lack of solid observational data on wolverines was hampering the development of effective science-based conservation strategies.

But observing wolverines in the wild has always been exceedingly difficult due to their relative scarcity, their secretive nature, and the remoteness of the habitat in which they are found.[8] Magoun herself had only completed the first study of wild wolverine families in 1985. Obviously, innovative strategies needed to be developed that would make it easier to obtain information on the behavior and habitat requirements of young wolverines.

In the summer of 2000, Magoun and two wolverine kits were dropped from a plane deep in the mountainous Alaskan wilderness, where she spent two months living in a tent, faithfully observing and documenting their behaviors as they transitioned from young wolverines to subadults. Her carefully trained scientist's eye provided extraordinary insights into the way the wolverines developed their skills to hunt and cache food, how they avoided predators and scent-marked their territory, and their patterns of movement across the terrain.

Just as important, by comparing her observations of wild wolverine kits with those previously done, she was able to demonstrate that data acquired in this more accessible manner could add valuable insights to the literature and aid in conservation plans for wild wolverine populations.

If Jeff wanted to mine all the major sources for the latest wolverine research, surely Audrey Magoun was the mother lode.

Judy Long of The Wolverine Foundation supplied him with the

necessary contact information to e-mail Audrey at her home base in Fairbanks, Alaska.

Luckily, it was an opportune time of the year to reach her. Magoun typically spent long months of every year studying wolverines in their natural habitat, sometimes remaining out of e-mail contact for weeks at a time, but by late spring she was just coming off the main season and back in the office when Jeff's introduction arrived.

As it turned out, Magoun had also already heard of the Michigan Thumb wolverine, and she had a few questions of her own.

What really drew her in was the success these amateurs were having with their trail cameras.

In mid-May, Jeff and Jason had acquired a new 8mm audio/video unit equipped with a motion sensor similar to the ones in their traditional still-image cameras. Each trip of the sensor triggered five minutes of recording. By the end of May, they had accumulated over 40 minutes of video footage.

Suddenly, the animal they'd been monitoring was no longer a static image. It was a three-dimensional, fully animated creature full of life and personality: climbing, digging, standing on its back legs to sniff the air for danger. They watched, mesmerized, as the wolverine came in close to smell the video lens, almost as if it were thrusting its nose into their own unseen faces. At one point in the exploration the wolverine became agitated, attacking some brush before standing on the back of the camera, causing the lens to tilt up.

It wasn't just the visual action; the audio recording gave the scenes a whole new level of reality too. They could hear the wolverine walking on the brush, the rough sounds of its claws on the tree bark, the noise of its powerful jaws crushing bone as it fed on the bait.

For Jeff, the switch to video marked the start of an entirely new phase in his relationship with the wolverine. The video camera was much more than a mechanical eye. It was a doorway, allowing him to step inside the wolverine's world and feel as if he were actually interacting with it. Slowly, bit by bit, its hidden, secret world was becoming as real to him as his own world at home and school.

It was this success with the videos that had caught Audrey Magoun's attention.

Jeff had written to Judy Long at The Wolverine Foundation, telling her about the footage, interested to know if any other researchers had had similar success. Judy had forwarded his query to Audrey, who wasted no time responding. She wrote this e-mail to Jeff on June 2, 2005.

> Judy copied your email to me regarding your success with remote video of the wolverine in Michigan. I was wondering if you are using a camera on the retail market or if you had one made up special. I used one made by a commercial company run with an 8mm movie camera on a carcass being fed on by wolverines. Even though it worked in practice sessions at home, it didn't work properly in the cold at the carcass site. Fortunately, I was there with my regular digital movie camera anyway, so got some good footage. However, I'd like to find a remote system that is dependable, even if it has to be special ordered.
>
> Congratulations on your work with the Michigan wolverine. Were you able to collect hair samples or scats for DNA analysis to determine the possible origin of the animal?

That was all it took. Kindred spirits, for sure! Jeff responded immediately, thrilled at the opportunity to swap tips and data with one of the leading experts in the field.

> Thank you for e-mailing me and I respect you tremendously for being so involved with the most interesting and amazing mammal in North America. The 8mm sensor unit I received from Woodland Outdoor Sports is an amazing unit and has worked flawless for me, whether be it cold or the elements Mother Nature has thrown at it.
>
> Within 2 weeks I had 40 minutes of video footage with audio, and now have over 2 hours of video footage. The president of Woodland OS Steve McNiel set it on 5-minute running time and the time is constantly running on the screen, along with the change in date each time a new day goes by. I am getting his night vision unit next, in an attempt to get some quality night footage.
>
> To see a *gulo* in action and get more than a glimpse of his athletic ability and agility, and tirelessness, to say the least has been incredible. It dwarfs any still picture I have taken of him. From one *gulo* lover to another, take care and good luck with future studies.

The pair immediately struck up an easy friendship, chatting via e-mail on an almost daily basis on their mutual favorite subject of conversation: wolverines.

"I was letting her know everything that was happening with my research, and she would in turn discuss what was going on in her research, which usually was accompanied by a few pictures," said Ford. "I could tell from our discussions the passion Audrey had for the wolverine species, *Gulo gulo*. And I really started to admire her tremendously with her intelligence, her knowledge and lifelong commitment to helping this species, which for Audrey was literally every day."

On her end, Audrey Magoun was intrigued by the mysteries surrounding this Michigan anomaly and curious about this gruff high school science teacher who'd championed the cause of answering those riddles. Despite Jeff's lack of credentials, Judy Long's introduction had bought him at least some measure of respectability and legitimacy.

Like a mouse unobtrusively sneaking into the house through the floorboards, Jeff had slipped beneath the radar into the inner circle of the wolverine world.

"Even with her immensely busy schedule, Audrey was always willing to take the time to answer the questions I had, and in great detail," he said. "Sometimes her e-mails back to me were pages long and often accompanied by pictures to drive home her point. Audrey's passion for the wolverine was a driving force that helped me to continue the research for so long."

Summer, 2005

✦ ✦ ✦

During June, the men had gotten some of their best pictures yet. But by the end of the month, as warmer weather arrived, they began to realize something was amiss. For the first time in over two months, the cameras took no pictures of the wolverine, only raccoons.

By early July, they had over 50 pictures of raccoons but not a single new shot of the wolverine. Jeff knew by now that whenever raccoon images were in abundance on their cameras the wolverine wasn't, and vice versa. Quite likely it was a matter of wise avoidance on the part of the raccoons, which knew better than to challenge an adult wolverine for food and only came around when they were confident that the larger, fiercer mammal was nowhere in the vicinity. This was the longest period of time they'd gone without a single wolverine sighting since that first photograph back in March. Obviously, the raccoons knew something they didn't.

Both Jeff and Audrey suspected breeding season had something to do with it. Male wolverines were known to cover huge distances to satisfy their biological imperatives. Although the Thumb wolverine's gender hadn't been confirmed, even a female might travel more now in search of a mate.

But just how far might that be? As far as anyone knew, this particular animal was well outside the range of any other wolverines. The closest known population was in northwestern Ontario, separated at this time of year by the impassible watery barrier of the Great Lakes. Breeding season could extend well into August. Jeff began to wonder how far it might roam in what would likely be a fruitless quest. Would his wolverine vanish forever?

Something else happened that month too. On July 24, 2005, Jeff received an e-mail from Audrey Magoun: it seemed someone had re-

The Wildlife Eye video system catches a photo of the wolverine's huge claws as it searches the research site for food.

ported a possible wolverine sighting near Ancaster, Ontario, about 150 miles east of Ubly, Michigan.

According to the report, the witness noted the animal had a reddish coloration to its fur, with black on the back and tail, remarkably similar to a wolverine's markings. The witness also described the animal's waddling gait as it moved away. Two similar sightings had also been reported in the same area, she said. Could it be Jeff's wolverine?

Jeff didn't think it was—he'd finally gotten new pictures of the wolverine on July 19. Despite its inexplicable disappearance, it was undoubtedly back and appeared to have settled in at home once more.

But the news was still exciting. If the reports were correct and there really was a wild wolverine near Ancaster, then it meant a population was closer than previously known. That made it even more feasible that his wolverine had crossed the ice on Lake Huron to reach the Thumb. Instead of having to make its way 500 miles around and across two

Great Lakes from the distant north, it would only have had to make what might have been a week's journey in wolverine terms.

Audrey was quick to caution that the Ancaster "sighting" might easily have been something other than a wolverine. After all, most sightings reported to The Wolverine Foundation turned out to be something else on closer investigation.

Jeff was keenly aware of the need for a healthy dose of skepticism when it came to eyewitness accounts. Each time his own pictures and articles were published, he was amazed at how many people called or wrote claiming to have seen wolverines in all kinds of unlikely locations and scenarios. One man insisted he had discovered a den on his back forty and that a male and female wolverine were hibernating there—something wolverines clearly never did.

Wolverine conspiracies abounded, complete with tales of ominous military types obscuring evidence like some wildlife version of aliens inside Area 51. One of the most bizarre accounts was from a Mayville woman who wrote to confide that just a few years ago her cousin's daughter had encountered a "scary-looking animal" lurking in a ditch as she waited for a school bus and had later positively identified it as a wolverine. Her cousin reported it to the DNR, she said, but was ignored. It was only after several more eyewitness reports of wolverines terrifying local children, she said, that the DNR swept in with the National Guard and loaded an entire pack of wolverines onto a helicopter, whisking them away to some remote location in extreme northern Michigan. Cousin Roy, she said, had been threatened by the authorities to never tell his story to anyone.

It was all pretty entertaining but also a little disturbing. Was he doing the right thing, publicizing the fact that this wolverine even existed?

It was fairly easy to ferret out the crackpots and crazies. He was also becoming more skeptical of others' methods and motivations even when it came in the guise of science.

Gregarious by nature, Jeff could never resist a query from anyone who wanted to talk wolverines. At the same time, however, he didn't think twice about slamming that door in someone's face if anything about the encounter annoyed him or raised a red flag. He was coming to think of himself as the wolverine's personal bodyguard as much as its publicity rep—and potential allies and associates could quickly find themselves kicked down to the rank of interloper or invader if Jeff's instincts told him that something didn't smell right.

Late that summer he was approached by Dr. Patrick Rusz of the Michigan Wildlife Conservancy, the controversial group at the center of the ongoing debate over the origin and status of cougars in Michigan.

According to Jeff, Rusz informed him that he and his colleagues had decided they'd like to learn more about Jeff's research and perhaps even set up a collaborative scientific study; in fact, they offered to underwrite it with a generous donation.

But it seemed to Jeff that their goal was to prove that wolverines had been living in Michigan undetected all along, just like cougars. Hunters in the Thumb region had probably never noticed any tracks prior to 2004, Rusz told him, because they simply hadn't known what they were looking at.

The notion offended Jeff and his partners. They were experienced, avid hunters and outdoorsmen who spent nearly every weekend in the woods throughout the fall and winter. A *gulo* track was huge and quite different from those of the other mammals common to the region. It was ludicrous to think they'd simply not noticed wolverine tracks before 2004—nor had anyone else, outdoorsmen or DNR officers alike.

Then Rusz asked that he be allowed to inspect some tracks in situ in order to confirm and document that what the men had claimed was in fact legitimate and not some kind of hoax.

At the DNR, wildlife biologist Arnie Karr had already inspected and verified their casts, and both Audrey Magoun and The Wolverine Foundation had confirmed their photographs. There might be a few people who didn't agree with what they were doing for various reasons, but not a single authority had questioned the honesty of their claims. Could anyone possibly think they were flying up to northern Ontario every weekend to grab photographs and evidence on the sly?

But the worst bad taste in Jeff's mouth was the notion that the wolverine's story should be taken out of outdoor magazines and into the scientific journals where it belonged.

Jeff had the utmost respect for bona fide research such as Audrey Magoun's northern boreal studies and Jeff Copeland's work in the West. He had no illusions about where his amateur efforts ranked compared to those of trained scientists performing serious academic research. He had a keen appreciation for the fact that it was only through well-documented, long-term studies like theirs that state, federal, and international wildlife management plans could find a foundation in solid scientific data. But at the same time, he felt confident that his kind of public outreach was just as valuable in fostering widespread interest in this little-known predator and creating popular support for wolverine conservation.

Who did these people think they were?

But Jeff was busy as a full-time teacher and father; he needed competent help to search for tracks and more fully document the wolverine's travels and habits. He needed someone with a real scientific background. In short, he needed someone like Rusz. Could he resist the temptation? It turned out he could.

Jeff finally agreed to supply Rusz with a detailed map of areas where he'd discovered wolverine tracks. Rusz could take it from there.

Rusz offered to pick up the map in person the following weekend. What he couldn't know was that Jeff had no intention of sharing anything with him.

In fact, the map was a work of pure imagination. Before Rusz's arrival, Jeff took out a copy of the *Michigan Atlas and Gazetteer* and began randomly marking remote areas throughout nearby Huron and Tuscola counties. Jason burst out laughing when Jeff showed him the result.

"You're an evil bastard, Ford," Jason said.

Combined with clues from Jeff's past published articles, Rusz zeroed in on the Tuscola State Game Area as the likely spot for the wolverine's home stomping grounds.

It was in reality 32 miles from the actual site.

"I don't know how long he looked in and around that area, but eventually he gave up the quest for the wolverine and his phone calls stopped and he went back to pursuing cougars and wild pigs," said Ford, chuckling.

What Rusz probably never suspected was that the answer was staring him in the face the whole time. He'd been the victim of Jeff's wicked sense of humor and irony.

"All he'd have to do was look at the map and go to the one county not marked on the map, and find a remote area in that county: the Minden City State Game Area in Sanilac County. It would have been easy had he known the lengths I would go to to make sure she was not disturbed."

Meanwhile, he had more pressing matters to attend to. Had he acquired any hair samples yet, Audrey asked. It was a gentle reminder of just how little he actually knew for certain about this animal he'd been studying for the past five months. In an e-mail on July 26, 2005, she wrote:

> How do you know that it is a "he"—without DNA (or wolverine in hand), one can't be sure. The small home range size may indicate otherwise. And the head size in some of the photos is small enough to be a female. How did you arrive at calling it a male?

He had to admit Audrey had a point. He'd only assumed it was a male based on speculations he'd made from the width of the tracks, which had measured 3.6 to 3.8 inches across. He'd been careful to only measure tracks that were a half inch deep or less to avoid the possibility of distortion. Compared to the measurements Judy Long had sent him back in March, this was on the large end even for a male.

What he hadn't realized, however, was that he'd been comparing the proverbial apples to oranges. The Wolverine Foundation's measurements, which he'd been using as a scale for comparison, were of direct foot measurements, pad to pad, on a captive wolverine. He was talking about tracks laid by a foot. Any impression in snow or mud would undoubtedly be larger than the foot itself, regardless of the depth of the impression.

The actual range for a typical wolverine *track*, Audrey said, was four to six inches from outside toe to outside toe. So rather than being large, she explained, Jeff's animal actually seemed rather small to be a full-grown male.

Still, some of the first photos taken of the wolverine when it had been treed in 2004 seemed to show a head size relatively large in proportion to the body, which would be more indicative of a male. On the other hand, many of Jeff's newer photos seemed to show a relatively small head, which would be more likely for a female or immature male.

Increasingly, Jeff felt compelled to get the DNA evidence needed to either connect the Thumb wolverine to the Ontario population or eliminate it as a possibility. And he didn't even know what sex it was yet!

Despite Jeff Copeland's polite refusal to get involved without the DNR's intervention, the Idaho researcher had given Jeff some useful ideas on how to proceed in collecting hair samples, and Audrey Magoun offered suggestions on how to fine-tune the technique.

It was best to get the samples before they sat too long in the field in the summer, Audrey said, because moisture and sun would eventually break down the DNA. If possible, put the snares in a shaded spot, she said, where they would be somewhat protected from the elements.

If he managed to collect some samples, she would send them to a lab that had already done DNA work on wolverines.

"I'm not sure of the cost at this point, but we'll work that out," she told him in an e-mail on August 13, 2005. "You can contribute what you think it is worth to you and I will pick up the rest since I am also interested in the results."

Jeff wasted no time putting the plan into action. Now that he had a clear sense of what needed to be done, he was more determined than ever to collect the physical data necessary to find hard-science-based answers to the many questions surrounding the Thumb wolverine.

His first hair trap was a confection of barbed wire wrapped around a tamarack tree in a spiral, each loop about four to six inches from the next. He hung bait high up in the tree so the wolverine would have to climb over the wire to get to it.

He then set the trail camera so its lens was pointed directly at the booby-trapped tree, about six feet away.

Right on cue, the camera dutifully recorded the wolverine climbing over the wire to get at the bait.

Unfortunately, the raccoons had done it too.

A close inspection of the hair trap revealed what appeared to be hair of two distinct types. Jeff excitedly collected each strand caught in the barbed wire, using tweezers to place them in a paper envelope, as Audrey had directed. She'd specified that the clumps from each barb should go into separate envelopes. But it was so windy, he simply stuffed them all in one envelope, worried he'd lose them otherwise. He was sure the lab technicians would easily be able to distinguish between them.

Then he carefully packaged them as Audrey had advised before shipping them to her in Fairbanks. He felt confident that he was closing in on the mystery's end now, all the answers to the questions that had been plaguing him only one mere lab report away.

Audrey Magoun knew the solution might not be that easy or elegant. She'd already reached out to a friend who was a wildlife biologist in the Department of Environment and Natural Resources for the Government of the Northwest Territories in Canada, seeking advice on a lab that might be able to provide a fast turnaround on some wolverine hair samples.

He recommended Dr. David Paetkau, president of Wildlife Genetics International, Inc., in Nelson, British Columbia. A species and gender analysis on a single sample could probably be turned around fairly quickly, he said, but speculating on a possible origin, especially for a sample from Michigan, could be challenging without an appropriate reference collection. Quite simply, they'd have to compare the sample against samples from a broader geographic area, and that was no small task.

Paetkau agreed. It would be straightforward to identify the species and sex, but it would take good local reference sets of 25 to 30 animals analyzed at 10 or more genetic markers to assign this individual to a specific locality. If they were willing to pay, he'd be happy to have a lab technician drop everything to turn around the species and sex identification in as little as 24 hours. But determining the locality? That was something that would need a good deal more discussion before he'd be willing to commit his lab's resources.

When she opened the package from Jeff on August 24, Audrey was disappointed to see so few hairs inside. Usually the labs liked to have at least seven good hairs with follicles attached, and she wasn't sure they had that. Then there was the question of contamination with raccoon hairs. The analysis would tell them if the hair was raccoon or wolverine and whether it was a male or female. It might even give them some clues on the origins of the wolverine, whether it might have come from Ontario or been a captive at one time with a genetic source from western Canada. But the cost of the analysis would be affected if they were unable to weed out the raccoon hairs first.

She decided to go ahead and send the meager samples on to Paetkau's lab, asking if he thought it worth the time and money to analyze what they'd collected so far.

Paetkau's reply was reassuring. The species test uses mitochondrial DNA, which is very abundant in every cell, so even one good hair follicle that hasn't been degraded by sun or rain could easily provide the required material for testing. He also estimated about a 70 percent chance of determining the gender from a single hair.

Meanwhile, Jeff and Jason were hard at work collecting more samples, refining their trap with more wire, and adding other adjustments that they hoped would allow them to snag more hair. It was the end of August now, and after mailing the last samples that week, they'd be pulling out again until January.

They managed to collect only three more samples, two of the hairs collected from the bark of the tree itself rather than the barbed wire. Again the camera showed both a wolverine and raccoons climbing the tree several times. Why was the trap so ineffective at snagging hairs? They resolved to try a different type of barbed wire when they returned to the task in January.

Meanwhile, Jeff had invested another 3,000 dollars in getting 500 digital video discs (DVDs) professionally produced from their best video footage. The load on the family credit card was growing, but he shrugged it off, confident it would somehow pay for itself in the long run.

Audrey Magoun was impressed, her respect for this amateur growing bit by bit. Could he get a picture displaying the animal's chest hair? Wolverine chest patterns are fairly distinct and never change, so it would be a helpful tool in distinguishing this animal from others should any similar "anomalies" show up in the Thumb.

Jeff could only wish such a thing would happen! Imagine what it would be like to have an actual mating pair of wolverines virtually in his own backyard!

As summer gave way to fall, Jeff busied himself with preparations for the approaching deer season. He was also using the wealth of still and video footage they'd acquired to make an increasing number of public presentations on the Thumb wolverine, mainly at schools and sportsmen's clubs. He was making a small amount of money selling DVDs and copies of photos, as well as collecting a fee at some events for his presentations. But it wasn't nearly enough to offset the investments of time and money.

Not all of the public response was uniformly positive either. Some people were concerned that Jeff's activities were compromising the wolverine's protected status, intruding too much on its natural state. Others worried that the animal might become injured on the barbed wire they were using for the hair-snagging traps.

He turned to Audrey Magoun for support, hoping to defuse any unwarranted criticism. While she couldn't supply any documentation, she could reassure him that she knew of several wolverines that continued to return to the same snag trees over and over, and none ever appeared to have any scratches or injuries from the barbs. In fact, she thought they actually learned over time how to avoid the barbs so that they became less and less effective at snagging hair over time.

By November, there was still no word from the wildlife lab in British Columbia. Audrey decided to give them a nudge. It turned out to be a wise move: DNA labs are notoriously slow, and somehow the sample had not made its way onto anyone's radar screen. Paetkau promised an answer by December 20.

Jeff had news of his own. On November 15, Steve Noble had been in the swamp deer hunting and found the wolverine's tracks following a snowshoe hare for nearly a quarter mile. On the same day, another deer hunter reported seeing the wolverine loitering beneath his tree stand about two hours after dawn.

The hunter's description of the encounter disturbed Jeff a little. When the wolverine first came in beneath the tree stand, the hunter said, it stood on its hind legs and looked directly up at him, so he knew it realized he was there. He'd also left his pack at the base of the tree, and had urinated there before climbing up, but none of these scents appeared to bother it. The wolverine simply looked around for a few minutes, then followed the hunter's tracks away from the tree stand for about 150 yards before veering off on a route of its own.

The encounter made Jeff worry that maybe some of his harsher critics had a point. Perhaps their ongoing presence had accustomed the wolverine to human scent and, even worse, that the bait traps might be causing it to associate humans with an easy source of food. He was relieved that gun season was almost over. What if the wolverine walked up to the wrong person? It could be shot dead before it had a chance to realize its mistake.

Audrey tried to reassure him. In an e-mail dated November 28, 2005, she wrote:

> In my experience wolverines are very smart and learn quickly. It is not surprising that this animal has learned to investigate areas where humans are to find food, but he may have done that anyway without your baits, because his association of humans with food may come from a broader experience than you are aware of. I share your concern because I'm sure there are some hunters who would be tempted to shoot a wolverine that comes that close—out of concern (unwarranted) for their own safety if for no other reason. I don't know what you can do about it, however, because using aversive conditioning has not worked well in experiments with wolverines that kill domestic sheep and I expect it would not work well in this situation either, especially if it is hungry. About the only thing I could suggest is putting out a large enough amount of food (dead deer) that it would be less inclined to wander over wide areas looking for food—although that is not a guarantee, because in an experiment we did with supplemental feeding of wolves to prevent them from killing caribou calves, the satiated wolves still went to the calving grounds—apparently hunger is not the only incentive for animals to investigate sources of food, as other research has shown. I've wondered if the reason the wolverine went missing for a while in the past was that it found a deer carcass at some distance from where you normally find him and it held him there for a while. If you can find road-killed deer and get it into an area during the hunting season where hunters are not likely to go, but the wolverine can locate it, then that is about the best you can hope for. Other than that, education about the wolverine and the fact that it is not a threat is the only other approach.
>
> However, it is possible the wolverine will stand its ground near a carcass and try to bluff people with growling, although it's very unlikely to go any further than that. Most wolverines however will run away if challenged; in fact, most wild wolverines would be gone before you could get close. Even though the wolverine knew the man was there, he might not associate a man in a tree the same way as a man on the ground, so it might be tree stand hunters that pose the greatest threat. What are your educational options working with hunters?

Jeff decided his best strategy was to write another article, this time fo-
cusing on the wolverine's nature and how unlikely it would be to attack a
human. In all his readings, he'd yet to find a single confirmed case of a
human killed by a wolverine. In fact, it was one of the most fascinating as-
pects he'd learned about the animal. They would confront wolves, bears,
and cougars, challenging animals many times their size over a carcass, yet
they had never been known to attack humans. Would that discretion end
up being this particular wolverine's fatal vulnerability?

The new, darker turn in the conversation caused Audrey Magoun to
pause and consider things that hadn't occurred to her until now: What
did the Michigan fish and game department think of what Jeff was doing? What was its opinion on the animal's presence in Michigan? Did
he have much contact with the state's wildlife officials?

Jeff's reply on November 28, 2005, put her at ease.

Arnie Karr, who initially took the pictures of him after the coyote
hunters treed him in February, 2004, has been in contact with me
regarding the wolverine. I have kept him abreast of the events as
they happen, including sending him pictures and my DVD, which
he responded in turn by sending me a disc with some of his pictures
he took of him. The overall feeling I get from them is one of appreciation for research they have neither had the time, money or manpower to commit to. You of all people, understand the commitment
and hours of work to research such an intelligent and elusive animal,
although it's been somewhat easier on me because the wolverine is
confined to a much smaller area than his normal home range would
be in areas of Alaska. Him sticking to 6,500 acres has helped.

The one request I did get from the DNR is when I resume my
research this winter, I am to use only bait that is natural in his habitat, such as deer, rabbit etc. No more chicken. They were very nice
about it and I understand their request. I get the impression that
they are eager to find out the sex and especially origin, so they have
given me the go-ahead to continue my research. I'm sure any biologist would find this Michigan wolverine unique, fascinating and exciting, just as I do. If a biologist doesn't find this animal intriguing,
then perhaps they have chosen the wrong field. But they have been
very supportive, as are many people around the state. I'm giving
presentations to outdoor organizations/sportsmen clubs around the

state, and many people are extremely excited about the DNA sampling, as we are.

It was enough, for the time being at least, to set aside any concerns she might have had about how state wildlife officials viewed Jeff's actions. But she was still worried enough to reach out again to the same Canadian wildlife biologist who'd connected them with the lab in British Columbia.

The biologist's response to her query on November 28 echoed the concerns that have become all too familiar to wildlife managers dealing with today's urban-wildlands interface.

> I'd say that if wolverines are successful in finding food (i.e. get into a kitchen and steal a roast) or find a source of gray water, they often return and may become increasingly determined to obtain further access to this reward or attractant. We've seen individuals become increasingly habituated, and may reside beneath buildings and chew through floors or walls to gain access. As with bears, these individuals can become more tolerant of human activity and may develop into "problem" animals and can potentially cause considerable property damage. Staff at mining camps have expressed concern about human safety.
>
> I'm not sure I'd use the word "aggressive" but they certainly may become bolder and perhaps venture into closer proximity of people and buildings during this search for food. This pattern is probably more evident in mid-winter (January–April) when wolverines are more nutritionally stressed. As with bears, I believe we need to encourage these camps to be proactive in terms of food, odor, and garbage management. As well, there's a need to use metal skirting around buildings and ensure that garage and kitchen doors are not left open.

Thankfully, they both were able to breathe a sigh of relief a few days later. It was November 30, the last day of firearm season, and Jeff had gone to the swamp to check his digital camera, which he'd placed over a deer carcass two weeks before. The wolverine had shown up promptly at 6:45 a.m., standing on top of the deer as the camera shutter clicked.

Another picture showed it in the background near the carcass's head. He was relieved to see his "furry friend" had made it safely through the firearm season. Muzzleloader season was still running, but he wasn't overly concerned about that, as very few muzzleloader hunters made it this far into the swamp.

It was Christmastime, and Jeff felt he'd received the best gift of all. His wolverine was alive and well and had made it through yet another year.

What still lay ahead for them was anyone's guess.

January, 2006

✦ ✦ ✦

Jeff was still posting to the website and doing updates for local publications, and by now, nearly a year after the story first broke, he was getting used to the notoriety.

Even Gianna Savoie, a producer from the popular PBS program *Nature*, had contacted him to discuss including his story in a science documentary being produced on wolverines. A year ago, few people outside the Thumb had ever even heard of Jeff Ford. Now it seemed all kinds of important people from all over the country were suddenly interested in talking to him.

One of those was an independent biologist who had made a name for himself a few years earlier working on genetic analysis of lynx-bobcat hybrids in northern Minnesota.

The encounter with Rusz had soured Jeff a little on opening up to any researchers except Audrey and her associates. But this scientist's research sounded fascinating, and his credentials seemed impeccable.

In an introductory e-mail, the researcher explained that he had read the media reports about the Thumb wolverine and was intrigued. He hoped to learn more about this latest wildlife anomaly and wondered if Jeff's group could collect scat that could be analyzed, as well as hair, assuring him that the costs for the DNA extraction would be absorbed by either him or the genetics lab at the US Forest Service's Rocky Mountain Research Station.

"Once the lab has an extraction, the DNA keeps for 30 years," the scientist explained in an e-mail dated December 20, 2005. "It's the processing of lots of genotype that adds up. We won't have to do a lot of that for now since we just need a genotype on the thumb wolverine for starters. I will have the biologist send a few vials."

That was good news, indeed. Jeff's credit card debt for the project

On hearing a noise, the wolverine instantly freezes and stares intently, occasionally standing up on its hind legs to get a better view.

was piling up well past $4,500—no small sum for a rural schoolteacher with a wife and two kids—and this was no time to start cutting corners. He needed to keep moving forward, whatever the cost, and the scientist's offer was tempting indeed.

In lay terms, a genotype is like a DNA fingerprint, which can be used to identify a specific animal and also to compare that animal's DNA makeup to others of its species in order to trace heredity and origins. While determining the genotype of the Thumb wolverine was now within Jeff's grasp, it was still only a first step in what could eventually be a lengthy and expensive process. Jeff welcomed anyone who might be willing to help fund or speed that process along.

Over the holiday break he did his best to oblige the Minnesota researcher, but finding scat turned out to be harder than he thought. He and Jason had acquired over 20 new pictures and plenty of fresh tracks in the week leading up to Christmas. But after splitting up and spending more than two hours scouring the snow around the area, they turned up no sign of wolverine droppings. They'd keep trying, he promised.

"He's hitting the venison hard now, and has to shit sometime, right?" Jeff joked in an e-mail on December 27, 2005.

Secretly, however, Jeff was getting a little apprehensive. He didn't want to be endlessly scouring the wolverine's territory, seeking scat and leaving his own scent all over the place. Right now they were very circumspect about their visits to the swamp, making their forays as brief and "clean" as possible. He was certain the wolverine had chosen this remote area as its sanctuary precisely because there was so little human presence. What this guy wanted them to do seemed incautious at best and might even force the animal to move off in search of a more isolated place where it felt safer from human incursion.

After the first couple weeks of communications, the researcher had also begun pressing for more specifics. Where did the wolverine spend the most time? Which areas were the most heavily populated with hares, the likeliest dietary staple?

They were reasonable queries, questions any researcher would ask. But ever since acquiring those first pictures back on the edge of spring in 2005, Jeff had worried that if he shared too much about the wolverine's habits and habitat, sharp-eyed observers would be able to use those clues to pinpoint its location beyond some vague region in the Thumb. But the last thing he wanted to do was make an enemy out of someone unnecessarily.

He'd unintentionally risked that already with Rusz by giving him the sham map. In fact, he'd followed it up by writing a scathing, heated letter in January to formally tell Rusz, once and for all, that he wanted no part of him or his organization's interest in a scientific study. It was only on Audrey's level-headed advice that he'd relented and set the letter aside.

More than anything else—even more than satisfying his own curiosity—he simply wanted to protect the wolverine from prying eyes, and the only way he could think of to do that was by safeguarding the exact location of its home territory. A secret is only as good as the first person who shares it. Three people—Jeff, Jason, and Steve—already knew the secret. He was determined that was as far as the knowledge would go.

Early on the three friends had made a pact that no one else would be let in on the location of their research site, no exceptions. Jeff had even shut Jason down when he'd asked to bring his dad to the bait site.

But it was tricky writing articles and giving interviews without divulging such basic details. Bolstered by the success of the phony map, he continued to give out a decoy location: Verona State Game Area in Huron County, directly north of Sanilac County. It was where coyote hunter Aaron Schenk had backtracked the wolverine's route the day after that first encounter in 2004, so it made perfect sense. Verona was close enough to be plausible but far enough away that casual explorers weren't likely to wander onto its tracks by accident.

The Minnesota scientist, of course, appeared to be exactly what he'd presented himself as: an interested wildlife researcher who meant the wolverine no harm. He also seemed to share Jeff's own pet theory: that the wolverine had crossed the ice of Lake Huron from Ontario to arrive in Michigan. But, despite his credentials and like-minded thinking, he was still a stranger, and Jeff was torn as to just how much he should share. Verona was an easy ruse.

But Jeff knew he couldn't maintain that deception forever. It was obvious they couldn't move any further in forming a partnership without giving him those basic details. Despite the lure of more assistance, Jeff backed away. He was back to working alone, outside the scientific community.

He turned his focus back to getting another sample to Audrey Magoun. By mid-January, there was still no word from British Columbia on the DNA results, and both he and Audrey were getting impatient.

They had hoped that by Christmas they would at least have the wolverine's sex confirmed. Now it was the first month of the New Year and they were still at square one.

As soon as the holidays were past, Jeff and Jason headed back to the field to check on the wolverine and fine-tune their camera setup. This time they weren't messing around: they hauled out a monstrous load of 50 pounds of venison scraps they'd accumulated from the recent whitetail hunting season, with three cameras trained on the tree with its new and improved hair-snagging trap.

On Saturday, January 7, they went back to find nearly all the meat gone—it seemed their *gulo* was living up to the meaning of the word *glutton.*

In reality, Audrey told them, the wolverine probably had just hurriedly stashed most of the scraps in nearby caches to protect it from other scavengers—a wise move on the wolverine's part. Audrey herself had once watched a flock of 30 ravens dispose of that much meat in less than 24 hours.

Interestingly, Scandinavian biologists have observed that wolverines can be quite judicious when it comes to food handling. When a carcass is deep in the forest, wolverines don't dispose of it as quickly as they would a carcass in a more exposed location. It seems they are clever enough to determine when the risk of competing scavengers makes it worth the time and energy to cache meat quickly.

Obviously, the Thumb wolverine didn't feel too confident about leaving all that valuable venison lying around. Was it worried about raccoons making off with the meat—or was it leery of the humans who had been leaving it?

All along, they had worried that their regular presence in the wolverine's territory might have a negative impact—either scaring it off or habituating it to their presence, which could put it in danger if it lost its natural fear of humans.

It only deepened Jeff's resolve to preserve the secret of the exact location of the wolverine's territory from anyone and everyone, regardless of their intent.

The bait might have been gone, but in its place the wolverine had left what appeared to be more than ample fair payment. Caught on one barb, clustered together like precious strands of gold, they found a clump of four hairs. At least one had a full follicle, maybe more.

What's more, the three cameras confirmed that the only animal that had been up that tree was the wolverine. Hopefully now they had enough for a real DNA analysis!

It's a good thing they had that to look forward to. On January 18, even as they were celebrating the newly acquired samples, word finally came back from the lab.

The first samples Jeff had sent weren't from a wolverine at all. They were all raccoon hairs.

Audrey wasn't all that surprised—when she'd first opened the envelope, the hairs had seemed a little different to her from those she'd collected herself over the years. But she'd hoped that they might have come from a lighter area on the wolverine's body.

Sadly, it had turned out her first instincts were right.

All these months of waiting, and still they were no closer to an answer than when they had started.

Jeff assured her that these latest samples were wolverine, guaranteed.

"We are not having ANY raccoons coming in this time of year, so all samples you are receiving are gulo," he wrote Audrey in an e-mail on January 19, 2006. "I'm keeping my camera on the tree, and it confirmed that no other animal has climbed the barbed wire tree."

Despite their initial high hopes, their latest collection of four hairs turned out to have only one strand with a complete follicle, so Jeff decided to hang onto it until they could collect enough hairs to make another DNA test worthwhile.

At least they'd managed to capture some good pictures of the wolverine's chest markings, which are used by researchers to distinguish one animal from another. It was a beautiful pattern, very symmetrical and distinct, and helped eliminate one possible answer to the mystery: the Thumb wolverine was definitely not the escaped captive from the Minnesota zoo.

Audrey assured him that as spring approached hair collection might become easier as the wolverine began shedding its winter coat. Jeff hoped so; their new wire trap had at least two dozen barbs but somehow the wolverine's coat had managed to pass through unscathed, time and again. He was at his wit's end on ways to improve it.

Even though spring might make hair collection easier, it also meant the return of the raccoons. So they decided they would leave the trap up and keep checking it regularly through the rest of the winter season.

Over the half year of steady correspondence between them, Jeff's trust in Audrey had deepened even as their mutual respect and friendship blossomed.

One pet peeve they shared was the unfortunate abundance of misinformation surrounding wolverines. Like wolves and hyenas, wolverines had long been the victims of bad public relations, portrayed as vicious marauders to be feared more than valued or admired. It was the kind of negative public image that made conservation efforts and research funding a tough sell.

For much of the school year, Jeff had been bringing the results of the wolverine project into his high school classroom, engaging his science students with this real-world example of amateur fieldwork right in their own backyard. Audrey applauded the effort, keenly aware that it was exactly this kind of engagement with young minds that could help elevate the wolverine in the public mind and make the species' conservation—indeed, all wildlife conservation—a priority.

There certainly was a need for that. She shared with Jeff her dismay over a set of wildlife "fact" cards that was currently being distributed across Canada. Titled "Nightmares of Nature," the set included one on wolverines that was littered with errors, exaggerations, and gross distortions of fact in the service of portraying the animal as a terrifying beast.

She wasn't the only one upset by the materials. She forwarded this Jan. 19, 2006, e-mail to Jeff from a contact at the Wildlife Conservation Society, an international nonprofit organization dedicated to funding wildlife research and conservation efforts worldwide.

> Sadly, this is the image of wolverines that we have to combat against. A series of postcards of animals were sent out to millions of folks across Canada (at least). It seems that they hit up subscribers of magazines. Here was their idea of the wolverine (nightmare of nature). I don't know if the same circulation hit U.S. I think The Wolverine Foundation should write them a letter about how they totally squandered this important education opportunity with misinformation about wolverines—Ironically, while we were sitting in a recovery team meeting today in Ontario, we discussed the myths such as those perpetuated here which plague this animal and how they impact wolverine conservation.

Audrey forwarded photos of the offending card to Jeff with her own e-mail decrying the disservice it was undoubtedly doing to the species to which she'd devoted her entire professional career. Her e-mail read:

> Many facts are wrong; for example, wolverines don't have retractable claws; no wolverine has been documented as weighing 70

pounds—that piece of information comes from someone mixing up kilograms and pounds years ago in a publication that has been repeated ever since. Killing moose, while it has occurred, is very rare and usually involves animals bogged down in snow or in poor condition. They certainly aren't vicious, any more so than any other carnivore trying to defend itself or its food. I question the bone-crushing ability as well, as the ones at Dale's [a fellow researcher] take a long time to crack open bones, just as my dogs do, but they probably do have stronger jaws than an equal-sized animal, which is what is unique about them. No one has ever tested their bone-crushing potential to my knowledge. I have always been amazed how dainty an eater they are—even when they are hungry—compared to my dogs that bolt their food. Other comments could be made as well about the information and image portrayed on the cards. I'm not trying to change the wolverine's image into something that it is not, but it's time the "facts" get squared with reality. If there is one thing that the wolverine has that seems greater than other species it's persistence, perseverance, and endurance but even that may not be unique within the mustelid family . . . we know so little about the behavior of these critters.

Incensed, Jeff vowed to make the fact card a new class project, assigning his students the task of ferreting out all the errors and correcting them, then sending a group letter to the company. Even if it did nothing to change the actual cards, he said, at least his students would go out into the world armed with this real-life lesson in not believing everything they read online or saw on TV.

Without Jeff consciously realizing the shift, his personal goals were evolving. It was no longer enough to passionately study and protect this one particular wolverine. Now he began to see what he was doing, in the classroom and through the media, as part of a larger crusade to promote research and conservation of an entire species—one that had gone largely ignored even among diehard eco-advocates.

Everyone wants to save pandas, elephants, and baby seals, right? Jeff Ford had discovered that his personal life's mission was to spread the word on wolverines. And, through his growing kinship to Audrey, he was also becoming an unofficial part of that larger wolverine community.

"Don't short-change yourself on the question of 'scientific' studies," wrote Audrey in a January 24, 2006, e-mail. "As far as I'm concerned, what you are doing is 'scientific' since all good science starts with 'observational' studies."

On Monday, January 30, 2006, Jeff wrote to tell Audrey they'd "hit the gulo jackpot."

Thursday Steve and I went in and Steve wanted to try something different so he rigged up duct tape and sticky strips over the barbed wire. Jason and I went back Saturday and discovered lots of hair. Three digital cameras and the video system all confirm that there was only Gulo at the site. There was one big clump in between the wires.

When you get the samples, in the clear cassette case, the bunch with the knotted end is that sample. The knotted end is the end that was stuck in the barbs, which after separating with my knife we were able to pull free, fully intact. That sample should be pristine. The many other hairs were pulled from the duct tape and one from the sticky strip. I didn't want to risk contaminating the pristine samples. The two envelopes were from previous outings. I'm confident we have enough for origin and gender, although we are going to continue to collect until you confirm we have enough. The hairs are quite large and again, must be Gulo because we have had no other animals at the site.

The other exciting news is we have our first night vision video of him. The system performed well, and the night vision of him is crystal clear. When we get enough footage, and get it burned to CD, I'll send you a copy.

Audrey was excited too. She promised to contact the lab immediately to find out their next move and see if she could pin down a timeline for the results. A Canadian scientist had been working with samples from Ontario wolverines, she told him, and they should be able to compare them with those from the Michigan wolverine. Wolverines from the Yukon and Alberta were the source for most captives, she explained.

"We may have to eventually contact some zoos in your region of the country to check on the source of their animals—just in case they have had escapes that they don't want to own up to," she wrote in her reply.

That same morning, Jeff mailed the hair samples to Audrey via three-day priority mail and sent Audrey an e-mail message:

> We're really excited about the prospects of finding out how this feller got here, so nailing down a timeline with the lab would be fantastic. Just out of curiosity, what's your gut feeling on how he arrived here? I won't print it, this is strictly off the record. I know how you scientific folks like to back all comments with evidence, but I just thought maybe I could get an opinion on it.
>
> A gentleman named Al Eisher, who is from the thumb and a movie producer, contacted me a while back and told me about a movie he produced in Alaska called *Running Free*. It is about a young boy who stumbles upon a wolverine kit in Alaska while on vacation and they become friends, and the wolverine actually ends up saving his life. It actually showed in Cass City, my current residence, in the 90s. (I must have been in Montana filming grizzlies, cause I don't remember it.) He sent me a copy and it has great footage. There were actually two kits they raised to make the movie, and the footage was shot in Alaska as they matured.
>
> Anyway, Al told me about a wolverine behavior as a prelude to mating called "lockjaw" where the wolverines hook jaws and roll down a hill. They showed that footage in the movie. It is really neat. I'm assuming you are familiar with this behavior, but is it an actual behavior displayed prior to mating, or just something they do while playing? I'm doing some presentations to high school students at different schools in the next week, and was thinking about including this, but wanted to make sure I'm telling the truth.

It was the kind of casual exchange that had become commonplace between the two of them. Audrey's lengthy reply was unguarded and candid, providing a rare insight into her personal opinions about the wolverine. In its own small way, it was also unspoken proof of the trust and mutual respect that had grown between them.

> The movie you mentioned was shown in Alaska too. The story is a little hokey but the footage is great. The behavior you referred to as "lockjaw" (I call it the "crocodile roll") is typical of wolverines from the time they are quite young; my kits did it all the time when they

were wrestling; I'm sure it probably does happen with mating wolverines that are playing around but it also happens in other circumstances, including when they are trying to get a piece of loose meat or hide pulled off a carcass, just like crocodiles do when they try to break off a piece of meat from a carcass before swallowing it. It probably also happens when they have serious fights but I've never seen that in a wolverine yet. But wild wolverines are sometimes found with wounds on the head and neck, which indicates that fighting occurs and probably involves this behavior.

I just heard some interesting news this morning. Wildlife survey pilots doing wolf surveys in Ontario spotted wolverine tracks near Chapleau, Ontario, which puts the nearest known wolverine within 300 miles of the Thumb as the crow flies. I've been hearing enough to convince me that it is at least possible that a wolverine could have come down from Ontario to the Thumb and managed to find a way to survive all the potential hazards. Just how he got there, I just don't know. I'm not too familiar with how built up the areas are around Michigan, even though I worked for a summer in the Seney National Wildlife Refuge, and what the ice conditions are like in the winter.

The DNA results won't be conclusive, of course, but it will sure help to make the argument that your wolverine is a disperser from Ontario if it turns out that the genetic make-up is closer to Ontario animals than Yukon animals. However, I'm afraid I'm not too optimistic that wolverines can survive long-term in Michigan no matter whether this is a disperser from Ontario or not. It just doesn't seem like enough habitat there.

But I'm hoping that at least this wolverine will have a long residence in your part of the world. There is a small group of isolated wolverines in Sweden (maybe 4 or 5 animals) that have been there for the past seven years and haven't spread out from there and haven't had young as far as anyone can tell. You might be able to have a situation like that in Michigan, but all the hunters around the area makes me skeptical about that possibility. Do you think the DNR would want to add wolverines to the Thumb? Have you talked to them about that?

She also fired off an e-mail to the wildlife genetics lab in British Columbia.

I believe we have now collected enough hair from the Michigan wolverine to try to do an analysis of the relatedness of this individual vs. wolverines from Ontario and western Canada. We have several clumps of hair, and a remote camera trained on the hair collection site indicates that the single wolverine was the only animal approaching and climbing the hair-snagging tree. [An Ontario geneticist] has indicated that he would be willing to help out by providing the information you need from the Ontario samples.

Could you please let me know how to proceed from here? . . . How many hairs with follicles would you ideally want to have for this analysis? How much would it cost to do this relatedness test as well as the sex of the animal? And how long would it take to do the test once you receive the material—especially if we were to pay for an expedited analysis?

There was more good news from Jeff too: his latest photos arrived in Audrey's office, and they were excellent. With the image's detail of the wolverine's chest pattern clearly evident, they'd have no trouble distinguishing this animal from any other.

Could they finally be close to the answers?

February, 2006

✦ ✦ ✦

Audrey Magoun's casual mention of wolverine tracks near Chapleau, Ontario, added fresh fuel to a fire that had been burning in Jeff's belly for nearly two years, virtually from the moment he'd first seen the Thumb wolverine on the evening news.

While he scoffed at the notion that wolverines—like cougars—had been living in Michigan undetected all along, he was equally annoyed by the DNR's dismissal of the wolverine as simply an escaped zoo captive or pet.

The previous year, wolverine sightings had been reported near Ancaster, Ontario, east of the Thumb and elsewhere in southern Ontario. Now tracks had been spotted 300 miles north, on or near the Chapleau Crown Game Preserve—a protected wildlife refuge northeast of Lake Superior offering 2,700 square miles of shelter to species as diverse as lynx, timber wolf, moose, otter, and mink.

As unlikely as it might be, was it really that unreasonable to think a lone wolverine might have found its way south across the ice?

Audrey cautioned him not to make too many leaps of logic: there simply wasn't enough solid data available to know for certain what was or wasn't possible when it comes to wolverines.

According to a 2004 research paper published in the journal *Northwest Science*,[1] only two ecological studies on wolverines had been completed to date in the lower United States, in northwest Montana and central Idaho. A species widely scattered over rugged, inaccessible terrain was difficult to study; hence, such research was rare. In fact, the insufficiency of data was one of the main challenges the US Fish and Wildlife Service faced in getting wolverines considered for potential listing as a threatened or endangered species.

Random reports notwithstanding, the Great Lakes region was sim-

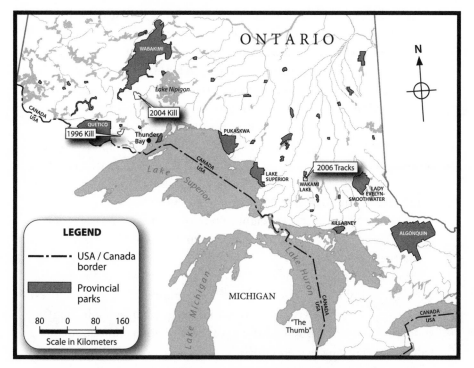

Map 3. To date there are only two confirmed wolverine reports in northern Ontario near the Great Lakes: a male wolverine trapped about 56 miles (90 km) west of Thunder Bay in November, 1996, and a male trapped in January, 2004, about 84 miles (135 km) north of Thunder Bay. Tracks were observed near Chapleau in 2006 about 250 miles (400 km) north of the Thumb. Courtesy of Neil Dawson, Wildlife Assessment Program Leader, Ontario Ministry of Natural Resources.

ply not considered suitable habitat for wolverines: it lacked the year-round cold and persistent spring snow cover believed necessary for their successful reproduction. Combined with the small number of historical records, those facts had led the Fish and Wildlife Service to conclude that wolverines were not and had likely never been full-time Great Lakes residents.[2]

Still, according to the *Northwest Science* article, in 2002 a young male wolverine (identified as M304) equipped with a VHF radio implant and GPS collar had covered a minimum of 543 miles in 42 days across the Wyoming and Idaho wilderness.[3] By the time the animal was legally harvested by a trapper on January 11, 2004, its greatest recorded distance of straight-line travel had been slightly more than 165 miles.

The boreal forest—known wolverine habitat—is a worldwide band of subarctic, conifer-dominated forest stretching across more than four billion acres of Scandinavia, Russia, Alaska and northern Canada. Ontario's portion of the boreal forest extends from the northern limits of the Great Lakes–St. Lawrence forest to the Hudson Bay lowlands, accounting for two-thirds of Ontario's forestland.[4] Chapleau—where bush pilots reported spotting wolverine tracks—lies at the southern gateway to that subarctic region, about 300 miles from the Minden City State Game Area and Jeff Ford's front door.

Jeff knew that Audrey was right to remind him to exercise caution in making assumptions and jumping to conclusions—but he couldn't help doing the math in his head.

Sure, he had no good explanation for why any self-respecting wolverine would be inclined to leave the comforts of the boreal forest to journey hundreds of miles south into relatively inhospitable, human-congested lands that offered little in the way of good climate, adequate habitat, or reproductive potential.

He had no hard evidence that such a journey had ever been made.

What he *did* have was a lone, highly improbable wolverine, alive and well and hundreds of miles from where every qualified expert said it should be.

Maybe an escaped captive was indeed the likeliest answer.

Or maybe not.

Increasingly, Jeff realized the puzzle's only solution lay inside the wolverine itself, locked in its DNA.

The ongoing conversation had also introduced a new and equally intriguing topic to the debate. Regardless of how the wolverine had gotten to the Minden swamp, it was there. Could it—or should it—be used as an opportunity to try to establish a breeding population? Once its gender was established, what were the chances that Michigan wildlife officials would have an interest in acquiring and importing a suitable mate?

Captive breeding, reintroduction, and other forms of human intervention have long been important tools in the conservation of threatened species. Of all Michigan's success stories, none has ever captured the public imagination more than the historic Moose Lift of 1985.

The project was headed by the state DNR working in coordination with the Ontario Ministry of Natural Resources and largely funded by the Michigan Involvement Committee of Safari Club International. Their mission: to capture, transport, and release a small population of wild moose into Michigan's Upper Peninsula.

On January 20, 1985, separate convoys of men and equipment set out from Michigan to remote Algonquin Provincial Park in northern Ontario. Deep in the heart of winter, the teams faced an extreme sub-arctic environment in which the windchill at times approached 100 degrees below zero. Using tranquilizing dart guns, they captured and loaded the sedated animals onto helicopter slings, airlifting them to a base camp 14 miles away.

Once at base camp, the animals were medically tested and fitted with radio tracking collars. Then they were loaded into shipping crates for a nonstop, 600-mile journey by truck to Marquette County in Michigan's Upper Peninsula.

Over a two-week time period, a total of 10 bulls and 19 cows were successfully transported to Marquette County. Enthusiastic crowds were waiting for them, ready to welcome these newest "immigrants" as they were released into the wilderness north of Lake Michigamme. The Moose Lift was deemed an enormously popular success. Two years later a similar effort released another 30 moose into the area.

Today, more than two decades later, studies have shown that the herd continues to grow at a rate of 5 to 10 percent a year and now ranges over 1,800 square miles.[5]

As dramatic as the Moose Lift was, its success as a reintroduction effort is dwarfed by sheer numbers in the return of Michigan's wild turkey population. Loss of habitat and unregulated hunting had wiped out Michigan's wild turkeys by the dawn of the 1900s. After four failed attempts, conservation officials purchased birds from Pennsylvania in 1954 and released them into the Allegan State Game Area. Within a decade, 2,000 turkeys were ranging free. Since then, aggressive habitat restoration programs by the state and the National Wild Turkey Federation have paid off in a statewide population now estimated at more than 200,000 birds—more than twice as many as are believed to have existed before European settlement.[6] Wild turkeys are now found in every county of the Lower Peninsula and most of the Upper Peninsula.

Not all of Michigan's wildlife reintroductions have been success stories, however.

In the mid-1960s, coho and chinook salmon were introduced to the Great Lakes and quickly became a hugely popular and highly valued aspect of the sport fishery. However, their continued existence in the Great Lakes was largely dependent on annual stream-stocking programs and today the Great Lakes salmon fishery has collapsed due to imbalances in the ecosystem triggered by invasive species.

And sometimes Mother Nature handles these matters on her own terms, without requiring a helping hand. There's no better example of that than the gray wolf, which managed to successfully reestablish itself in Michigan after human reintroduction attempts had miserably failed.

Around the world, the lesson seems to be that reintroduction efforts are largely fruitless without vital factors such as sufficient genetic stock, adequate protected habitat, and long-term funding and commitment to support the species' ongoing survival.

That was hardly the case for a single, anomalous wolverine living in a Michigan swamp.

Still, Jeff couldn't help but be excited at the notion of introducing another wolverine to the Thumb in hopes of establishing a breeding pair. He was even willing to put up 1,000 dollars of his own money toward that goal.

In a January 31, 2006, e-mail, Audrey Magoun expressed empathy for Jeff's dream and even acknowledged that there were means available to acquire a suitable mate. But the odds of that happening—or if it did, having it actually turn out to be a success? Highly unlikely.

Still, she couldn't help closing her reply without adding a wink and a whimsical sigh of regret.

I'm not sure what it would cost to bring a wolverine to Michigan; we might be able to talk Ontario into letting Michigan have one since they have had some trappers unhappy with wolverines on their lines and they are not allowed to trap them, so some that accidentally get caught anyway they have to turn the animal over to DNR. If they were to catch one alive and could hold it to remove a troublesome animal, the DNR might be willing to give it to Michigan. It would take some negotiation between Michigan DNR and Ontario DNR.

I think the biggest hurdle would be trying to convince Michigan DNR that there would be any point to it since there is so little habitat available in Michigan for expansion of a population and there may be people in Michigan who wouldn't want more there. It would certainly be an interesting political discussion but I wouldn't be too optimistic about the prospects. Even Colorado thinks it's too hot a topic to bring up—but Michigan is one wolverine ahead of them and has proof the wolverines live a quiet uneventful life! And after all, it's the state animal, right?

Jeff was fairly confident that the Minden City bog could support another wolverine. Even though it was true he'd been supplying food for the past year, the wolverine had obviously survived on its own for a year prior to that, maybe even longer. A respected local trapper, Dale Abbott, told Jeff he'd run into tracks back in the late winter of 2003 in a region of the Minden City bog known as the Ross Hills. Every February and March, Abbott was back in the bog trapping beaver along the canals, so he was keenly familiar with the habitat and anything unusual was bound to catch the attention of his seasoned eye. He'd noted the strange tracks weaving in and out around six beaver dams and suspected the animal that had left them was hunting beaver.

At first, Abbott assumed a bear was the most likely culprit. It was only after he learned of the wolverine's presence that he suspected that was what was probably stalking his beavers.

If Jeff's dispersal theory was right, the 2003 appearance made perfect sense. That winter, record-breaking cold threatened to freeze Lake Superior for the first time in more than two decades, with ice covering more than 90 percent of its surface according to an Associated Press article dated March 6, 2003.[7]

Satellite images taken from the International Space Station on March 12, 2003, showed large sheets of ice almost completely covering Lake Superior, clogging the Great Lakes' shipping lanes with ice and necessitating the use of ice-breaking ships to open the lanes for the beginning of the shipping season. The ice on Lake Superior was up to two feet thick in some places.[8]

Of course, no one had connected those dots at the time; back in 2004, everyone had simply assumed the encounter with the coyote hunters had marked the wolverine's arrival. But what if it had actually arrived a year earlier—not on a garbage truck but across the frozen Great Lakes ice?

No matter which direction his mind traveled, it always seemed to circle back to this, didn't it?

Where were those hairs with the DNA?

It turned out they were en route to a genetics lab in the west. On February 4, 2006, Audrey Magoun notified the lab that she was sending three or four packs of hair in separate clumps. Most contained only one or two hairs, and some had broken follicles. However, there was one clump with about 10 hairs, which hopefully would provide enough DNA to do an analysis to determine where the Michigan wolverine had come from.

She also alerted them to the possibility of contamination in the samples. There was a tiny bit of foreign matter on the hairs in the largest clump—some on the center of several hairs, sticking them together. There was also a very tiny spot of pink on one hair where the follicle should have been. She was worried this could be blood from something other than the wolverine—perhaps deer or something else on which the wolverine was feeding.

"My best guess on the source of the wolverine is either an escaped captive or a wild disperser from Ontario, the nearest known extant population although I wouldn't be surprised if there are a few wolverines in northern Minnesota now since the population in Ontario seems to be expanding, but those should still be closely related to Ontario animals," she wrote. "If it is a captive, the most likely genetically related individuals would be from the Yukon or Alberta, where most wolverines in captivity have come from, but Alaska is also a possibility because I sent a wolverine to Minnesota years ago, but I'm not sure that one ever bred. I will try to find out from St. Felicien Zoo in Quebec the source of their animals and if any have escaped. Supposedly an escaped wolverine was trapped in Quebec last winter."

She also offered to send along some hairs she had in the freezer, taken from a number of animals in Ontario, as well as tissue samples from wolverines in Ontario, the Yukon, and Alaska.

Audrey was doing all she could to get answers for Jeff, but he couldn't help feeling a little anxious. It had already been two years since the wolverine was first sighted. Wolverines typically stay with their mothers for one or two years, so it was quite likely the Thumb wolverine was four years old at the youngest. But it could be far older than that—he had no way of knowing. Audrey had told him that wolverines in the wild had a life expectancy of 10 years at best, although animals in captivity had lived as long as 20 years. How many years did his wolverine have left?

As far as he could tell, the wolverine covered a large territory but traveled mostly in circles, and it appeared to stick pretty tight to the 6,500-acre bog with its miles-long maze of canals dug in the 1950s for waterfowl. But what if something happened to Jeff and the free meals stopped? Would it wander farther and out onto a road in the path of an oncoming car? What if it was pressured by dogs again or caught in a trap?

Trapping was his greatest fear, but he felt powerless to do anything against it. If he informed the public of the wolverine's location in an effort to persuade people not to trap in the area, someone might use the information to try to kill it instead. He knew not everyone in the area was a fan. There had even been an anonymous letter published in the *Cass City Chronicle* in which the writer vowed to "trap the wolverine and save the children of the thumb from the wrath of the ferocious wolverine."

While he wasn't a trapper himself, he knew and respected many trappers and fully acknowledged the legality of their sport, just as he wanted others to respect his love of bow hunting. He also knew that the letter writer hardly represented trappers as a whole and that most wouldn't intentionally cause the wolverine harm. But a trap makes no moral judgment about the creatures caught in its steel jaws. The issue now was keeping the wolverine safe—and no trapper's rights trumped that in his mind. He vowed that if he found any traps anywhere in the area he'd come to call the Sanctuary, he'd spring them and take them out.

While Jason and Steve still helped out with the research, neither seemed as concerned as Jeff was about potential risks to the wolverine's well-being. In that regard, it had become his obsession and his alone.

Whenever he had free time within a few days of a fresh snow, he'd drive out to the access areas around the swamp to check for fresh tracks and traps. He hadn't found any yet, but he was painfully aware that one person trying to police 6,500 acres of swamp, trees, and brush wasn't much of a surveillance force.

He was absolutely convinced the threat was real. The previous weekend he'd followed a set of human tracks that crossed the wolverine's trail twice and came within 70 yards of their cameras and video system.

Audrey tried to calm him down in this e-mail on February 17, 2006.

I'm not sure it would be a good idea to remove traps because trappers get pretty upset and may try to discover why their traps were being messed around with, and that could cause unwanted backlash: your cameras treated the same way, discovery of what you are doing there, forcing [the DNR] or wardens to get involved, and I would think you would want to keep good relationships with them.

The only traps that would probably be effective on wolverine are coyote snares or the larger leg-holds or 220 or 330 conibears. Have you discussed this worry with [DNR] folks who are aware of your activities? Perhaps they know the trapper and can tell you if he is a reasonable guy who would pull any traps dangerous to the wolverine? Believe it or not, some trappers are really nice people who may be very interested in keeping the wolverine alive. But I understand your concern about letting other people find out because word would spread and there are the other sort of trappers out there as well. Hopefully, you will not be faced with the decision.

But very quickly Jeff found he had far more immediate worries than any hypothetical traps. That Friday he and Steve went out to check the cameras and found about two-thirds of the chicken they'd put out six days before still untouched. In 40 trips since they'd begun supplementing the bait sites with chicken, it was the first time the wolverine had left any behind.

Photographs taken on February 12 and 15 showed the wolverine climbing only two feet up the first bait tree before dropping back down, leaving the chicken untouched. It left a dozen scraps at the second tree and only took a few pieces from a third.

Was the wolverine sick? Not long before, Jeff and Jason had packed in a doe carcass they'd found as road kill. They knew it was less than 24 hours dead, but when he gutted and skinned it that night, he found the car had struck it so hard that the impact had burst the intestinal sack and the entire animal had a foul smell.

Since wolverines are carrion eaters, they'd assumed it was still alright and hauled it out to the site. They had video of the wolverine feeding on it shortly after. Could the carcass have been contaminated and now the wolverine was sick? Was its digestive system equipped to handle such things? Jeff knew the DNR had told them to quit using chicken; they'd only continued because it was so difficult to keep up a steady supply of wild game, and it was the only way they could ensure getting the hair samples needed for DNA testing. Now, ironically enough, after all his concerns about risks from others, it seemed all too possible that he had harmed the wolverine himself, and with the wild game that was supposed to be its natural food.

Or was there something else going on?

On the way in that day, they'd come across three men with backpacks pulling covered sleds, heading into Minden toward the south end of the canals. They'd claimed they were camping, but Jeff was suspicious: the windchill overnight was predicted to be 10 degrees Fahrenheit. That seemed a bit cold for a camping trip.

Jeff returned a day later to find that three more trucks had arrived. He also noted that the DNR gate had been opened, with tracks indicating that a fifth truck had driven in via that route—an unusual amount of official activity for this time of year. At the bait site, he saw that the wolverine had come in but had again refused to touch the chicken or the remaining deer carcass.

Could it be the DNR was trying to collect photos or hair samples of its own? Or check up on Jeff and his buddies to make sure they were not using chicken as promised?

Or was it someone else? Three local papers in the past month had run stories on the team's efforts to get hair samples to test for gender and origin. Jeff had also recently done a radio interview that was broadcast on at least 14 stations across the state. He'd been talking about it at presentations to high schools and sportsmen's clubs and had posted it on the website. He'd even talked personally to the local conservation officer for a full hour the previous weekend, sharing details of their research.

Jason had also talked to some men who owned hunting property nearby, on the west side of the game area. They'd told Jason that something, possibly a bear, had ripped up a downed tree on their land and they'd been baiting and setting up trail cameras in the hopes of getting a picture.

They'd made no mention of a wolverine, so Jason had dropped it at that, afraid if he probed any further it would arouse suspicion.

If all that were true, Audrey reasoned, then it was quite likely that the wolverine wasn't sick at all. It was simply a little too well fed. On February 19, 2006, she wrote:

> It is so close for a wolverine to where you are operating that he is likely to be traveling back and forth between the sites; perhaps doing a lot of caching at the new site and just visiting yours to make sure his caches and food sources are still there. It's not encouraging news, because if someone else is baiting for him, he will end up there.

A month ago, if someone had suggested another group was setting up camera traps in the area, Jeff would have come unglued. But at that particular moment in time, the only thing he felt was relief.

"Thanks for all your time and insight, even during my 'paranoid' times," he wrote to Audrey in reply. "I just care for the furry little feller."

March had arrived, with another spring fast on its way. And against all odds, the Thumb wolverine still survived, and remained.

Spring, 2006

✦ ✦ ✦

It had now been a full year since Jeff and his buddies had captured their first photos of the wolverine. But while Jeff's passion for wolverine watching had blossomed over the months into a full-blown obsession, Jason and Steve found their enthusiasm on the wane.

On March 29, Jason sent an e-mail to Jeff that marked the slow beginning of the other partners pulling away.

> I think we need to get on a two-week schedule for checking cameras. Working six days and trying to get things done around the house is getting to be too much. With Melissa going to school next year, I need to get all my [deer] scouting and stands placed this year because next year I can't afford to be running all the time. It looks like this will be my last hunting season for two years 'til Melissa is all done. She is going to be working and going to school so I am going to have to be doing everything I can to make sure she gets enough study time. Starting next year she will be working every weekend for the year . . .
>
> Saturday I will bring the turkey and the camera. That will put my pack at about 30 pounds. It's up to you if you want to bring some big pieces of venison. Also after the Elkton presentation, I don't want to do any more. It just takes too much time and I don't get much time at home. Jordan and Morgan start baseball in a couple of weeks.

Jeff wasn't surprised. He'd known all along that his interest in the wolverine had bordered on mania—and if he didn't, his wife and family had no trouble pointing it out.

How many times had he canceled on family get-togethers because he "had to go out to the swamp for research"? His wife Amy knew the whole litany by heart. He had to collect the hair samples before the

DNA deteriorated and became unusable. If he didn't check the camera batteries, he might miss that one moment he'd been hoping to capture on film. He had to restock the bait supply or risk losing the wolverine's interest.

It didn't matter if these were excuses or a valid rationale; the end result was the same. Weekend after weekend, she was left to fend for herself with the kids.

Even Jeff's parents were starting to get concerned. They gently suggested that maybe his obsessive level of research should be "toned down a bit."

Jeff's reaction was angry and defiant.

So he was supposed to just walk away? Give it all up after all this expense and time? The rarest mammal in North America walking around the Thumb, practically in his own backyard, and he should just quit? To hell with that. To hell with everyone.

Then his dad took a different approach. Didn't Jeff realize his camera baiting might be fostering in the wolverine a dependence on humans for a steady food supply? That could reduce its natural instinct to avoid the scent and presence of humans and even encourage it to seek out human habitations as a source of food. Jeff should consider that he might ultimately end up being the indirect cause of the wolverine's death.

The words hit him like a fist in the gut, going deeper than any of Amy's quiet complaints ever could. There was no one on earth Jeff respected more than his father.

As a teen fresh out of high school, Jac Ford had started out sweeping floors at General Motors' Saginaw Steering Gear. Despite his lack of a college education, he stubbornly worked his way up through the ranks to become superintendent of quality control at two of the plants, with thousands of employees under him. Two years before retiring, he enrolled in college classes at Central Michigan University, earning a degree in business management for no other reason than the simple personal challenge and satisfaction of earning it.

In the years following his wife's tragic death, Jac Ford had remained fiercely devoted to raising his two children even while working long hours. He'd taught Jeff and his stepbrothers everything he knew about hunting and fishing, instilling in them a love and passion for the outdoors as only a father can.

When Jeff was eight years old, his father had taken him fly-fishing on the Yellowstone River in Montana. Fearing the boy was too small and young to withstand the current without being washed away, Jac would put his son on his shoulders and wade out into the rushing waters so Jeff could cast to the rising cutthroat trout.

As a retiree, Jac had gone on to become a professional fly-fishing guide in Michigan and Montana, earning a reputation as being among the top guides in the Midwest. At 70 years old, he was still guiding every spring and fall, getting up at 5:00 a.m. and often not coming off the river until seven or eight at night.

Jac had always been there to lend his support and wisdom, and there were few decisions in life in which Jeff didn't heed his father's counsel. Jeff had so much respect for his dad that he had never talked back to him, never challenged him. For Jeff, Jac Ford's opinion was the final word on any subject.

But now here was his dad calmly questioning the wisdom of his son's ongoing actions with the wolverine.

Yes, he had to concede that his father had a point. The feeding could undeniably put the wolverine at greater risk in its assimilation of humans and their food.

But he also believed the weekly baiting was reducing its need for excursions outside the protected cover of the swamp. Wasn't that much less dangerous than if it were venturing out in search of food across busy roads, near farms and private woodlots?

Beyond every conceivable argument, pro or con, he had to admit he had one undeniable motivation that trumped everything else. Right or wrong, he knew in his gut he wasn't supplying food simply as bait for research purposes. He was making sure it had what it needed to stay alive.

He had to know that the wolverine was okay, no matter what anyone else thought or said. He had to.

And that was something about which not even Jac Ford could change his mind.

On the same day that Jeff received Jason's e-mail about cutting back on his participation in the wolverine project, other news was on its way that would do nothing to put the brakes on Jeff's own continued dedication. In fact, it would have the opposite effect, propelling him into a near-obsessive overdrive.

On March 29, 2006, Audrey Magoun received the results from the genetics lab on the second set of three hair samples and forwarded them on to Jeff as she'd promised.

This time Jeff and his buddies had done their work right: the hairs were indeed all wolverine. At the lab DNA extractions were performed on all three samples, analyzing the control region of mitochondrial DNA (mtDNA) that had been previously used to evaluate wolverine populations in four other studies. All three hair samples matched Haplotype C, a genetic marker that has thus far been found only in the Alaskan Kenai Peninsula and southern Alaska. The haplotype is fairly common in those populations—detected in 38.5 percent of the wolverines tested—and has not been reported anywhere else.

The most plausible explanation was that the animal had been transplanted from southern Alaska or the Kenai Peninsula; therefore, the wolverine was probably a released or escaped captive, just as Michigan's wildlife biologists had first speculated. In all likelihood, this was not a natural disperser that had made its own way south from Ontario.

It wasn't an absolute conclusion, however. The geneticist was careful to point out that they couldn't entirely rule out Ontario since the lab didn't have data from eastern Canada. Without testing eastern Canadian populations for Haplotype C, they explained, they couldn't say with certainty that this particular haplotype couldn't also be found in Ontario. Still, it seemed unlikely: researchers had never detected Haplotype C east of southern Alaska in any of the sampling that had been done across a wide swath of the Rocky Mountains and Plains regions.

The news sent a shock wave coursing through Jeff's system. He'd been waiting for this for so long it was hard to believe the day for some answers might finally have arrived.

He wasn't the only one interested in the results. One of the researchers who'd been rebuffed by Jeff the previous year had caught wind that DNA testing was under way and called Audrey Magoun directly, seeking the results.

She immediately alerted the lab, asking it not to release the information to anyone before Jeff Ford and The Wolverine Foundation had the opportunity to publish it in the popular media. The lab agreed.

Jeff was starting to understand that wildlife genetics, like many cutting-edge sciences, was a highly competitive field where researchers often raced each other to publish results and duels between rival theories were fought not on some athletic field or battleground but within the pages of scientific journals.

The lab's careful wording and unwillingness to speculate more than the data at hand could definitively quantify also made him appreciate just how exacting those findings had to be before any theories were allowed to see the light of day.

That should have cautioned Jeff about how circumspect he needed to be when dealing with the media. Unfortunately, it was too late for that.

In an article published in the April 2006 issue of *Woods-N-Water News*—on the newsstands before the lab's letter arrived—Jeff had made no bones about sharing his own theories and opinions on the matter, regardless of whose feathers might get ruffled.

Another interesting piece of information was brought to my attention by Audrey Magoun a few weeks ago. In January, 2006 a research team studying wolves came across fresh wolverine tracks near Chapleau, Ontario. This area is much further south than a wolverine's range was thought to be. In fact, the area is only 90 miles north/ northeast of Sault Ste. Marie. There also were two sightings last winter near Ancaster, Ontario, which is only 250 miles from the thumb.

The significance of these events is the wolverine range is much closer to our area than originally thought, and therefore it's much more likely that this wolverine did indeed cross the ice of Lake Huron to arrive here.

Wolves have now been verified crossing the ice into the Lower Peninsula, and wolverines have the ability to travel much farther than a wolf. A prime example is a wolverine that was GPS radio collared in the Greater Yellowstone Research study in 2002. In May of that year he traveled 140 miles in 7 days. His total movement over 42 days was 543 miles, all while climbing over mountain peaks. That's some serious movement. It is quite conceivable that this wolverine now living in our Thumb of Michigan took a road trip from Ontario. DNA analysis will tell more.

By the time the article was published, of course, the evidence appeared to be leaning toward the exact opposite conclusion, but there was no taking it back now. On March 30, 2006, Magoun sent this e-mail to Jeff.

Well, Jeff, with the results on the DNA, it looks like the Michigan wolverine has a good chance of being an escapee—our next step would be to be sure that the Ontario wolverines don't have the same haplotype found in the Michigan wolverine—that would clinch the conclusion that the Michigan wolverine is most likely an escapee—perhaps the one from Minnesota. I'll try to find out more about any escapees.

Undaunted, Jeff was just excited to keep probing deeper into the mystery, wherever it might lead. Jeff's reply is dated March 31, 2006.

Interesting!!! How long would it take to compare to the Ontario DNA? Was there any DNA from your previous studies that it has already been compared to, or is the existence of the type C at 38.5% pretty strong evidence that he came from Southern Alaska?

How many confirmed escapees have there been in Minnesota? More than the one caught on tape in the town of Zumbrota? I don't think it could be that wolverine. He was caught on tape 12 days before the coyote hunters ran him in 2004, and that's a 500-mile trip from there, including crossing lake Michigan and navigating around many major cities.

How many wolverines did you give to Minnesota? Fascinating that the Michigan Gulo might have ties to you. What a story!! I'm keeping everything hush-hush until we get some more answers, and I'll keep the report secure. They didn't say how long on the gender. Did they mention that to you?

Audrey's reply, on March 31, 2006, further speculated on the renewed possibility that the Thumb wolverine might have escaped from a zoo.

I have to discuss the Ontario DNA with them. First, we have to get a source. I have some hair frozen from 2 years ago but I don't know if that will be good enough. I don't know how a wolverine from SE Alaska would have ended up in zoos unless it came from animals caught years ago in Cordova and I don't know where those animals were sent. Have to chase this all down.

I only sent one to Minnesota—from northern Alaska—and I rather doubt that one ever bred and should be long dead by now, I would think.

They should know the sex of your animal very soon. I'll call the first of next week if I don't hear from them today. Meanwhile, I have someone from the Minnesota Zoo sending me info on their wolverines and possible escapees and origin. Do you know who I might contact at the Michigan zoo? Have you talked to them about possible escapees?

She followed that e-mail with the latest update on the Minnesota escapee that had so long hovered around the edges of the Thumb wolverine's tale.

Apparently the wolverine that escaped from the Minnesota Zoo was a female and she escaped May 13, 2003. That would be plenty of time to reach Michigan if the one caught on tape in Zumbrota (where is that?) was not the one run up the tree by dogs in Michigan. Once we find out the sex of the one in the Thumb, we will be able to straighten some of this out—perhaps. If it turns out to be a female, it could belong to the Minnesota Zoo and they may want her back. How do you feel about that? She has an identity chip so they will know if it is her.

Two Thumb wolverines? To Jeff's mind, that was highly unlikely given all the time he'd spent studying the animal. If it took adding a second wolverine to the equation to keep the Minnesota zoo escapee in the running as a potential answer for the origin riddle, then the theory had even less credibility than he'd thought.

His April 2 e-mail to Audrey included these thoughts as he prepared to write another article.

With all factors being considered, such as the lack of DNA from Eastern Canada and the large missing space of DNA between Alaska and myself, what kind of descriptive terms or %s could be assigned to the lab results? I'm thinking in terms of accuracy for my article. Would "almost impossible s/he came from areas other than SE Alaska or the Kenai" be too strong, or would "probably from that area" be a better description, or something else?

I would like to include a quote from you on the lab results, or anything else you feel pertinent, but understand if you decline.

How old was the Minnesota Gulo when she escaped May of 2003?

On April 4, 2006, a zoologist at the Minnesota Zoological Gardens responded to Audrey's request for information. The information was enlightening in that it finally laid to rest one of the most persistent theories on the Thumb wolverine's origins. At the same time, it only served to deepen the mystery, making it seem more hopeless than ever that they would ever find a satisfactory answer.

Your inquiries about escaped zoo wolverines finally made it to me. We did have a female that disappeared from the Minnesota Zoo in 2003 (see below, and attached pictures).

It was particularly disappointing because she was a British Columbia animal, whereas most of the zoo population are of Yukon origin. She had a transponder in her, though I would have to do some digging to find the number. Known as Tika, she was a true trickster; getting to places in the exhibit she wasn't supposed to get to, continually overcoming our efforts to prevent that, getting out of cages that had held numerous wolverines over many years, escaping to the keeper hallway to eat dog food—then returning to her cage, and finally just disappearing.

Once she'd studied the photographs, Audrey wasted no time in contacting Jeff.

Here is some very interesting news and in my opinion verifies that the Minnesota female that escaped is not the one in Michigan. Compared these photos to the nice photos you have of the chest pattern on your animal.

The wolverine's long, nonretractable claws and powerful front legs help it to easily climb trees, dig holes, and catch prey.

Although it has the same type of patterning—that is, a nice "necklace" of white markings, the markings are not identical and also, Tika seems to be a somewhat darker colored wolverine.

So the mystery remains where your wolverine came from and my guess is that we may never find out unless someone "fesses up" to losing one, and can prove the Michigan animal is the same one.

Jeff, meanwhile, was ready to roll out another big story in *Woods-N-Water News*, unveiling the findings from the genetics lab that contradicted his earlier theories. The publisher's deadline was fast approaching. Was there any way they could get the gender results in time?

Jeff had no illusions that the articles he was writing were on a par with legitimate research papers published by scientists such as Audrey Magoun and Jeff Copeland. His audience was typical hunters and outdoorsmen, laypeople for whom terms like *haplotype* were a foreign language. But at the same time, he took fierce pride in what he was doing.

"Although research studies resulting in scientific dissertations are essential at persuading governments to justify protection of an endangered animal, rallying the common man I feel is also essential in the gulo world," he wrote in an April 3 e-mail to Audrey.

He decided to write the genetics lab directly, asking for some quotes for his upcoming story. "At this point," he wrote, "prior to checking the Ontario samples, if you were to assign a % to confirmation that the wolverine came from the Southern Alaskan area or Kenai, what would it be? Can you give a description of what 'Haplotype C' means?"

A lab scientist responded with a lengthy reply, taking great pains to explain the complex subject of wildlife genetics in a way that anyone could understand. Not surprisingly, that proved to be more of a challenge than the scientist might have thought.

To answer your first questions: To assign a probability we would need to ask a very specific hypothesis such as, "what is the probability this sample originated from a Southern Alaska population versus an Ontario population?" We can't ask this question given that we don't have any Ontario samples to compare with. We can state that despite the hundreds of samples examined east of Alaska, this "haplotype" has not been seen. Thus it would be very unusual, but not impossible, to have a distinct distribution of a particular haplotype.

Your second question was about this Haplotype C. A haplotype is basically a form of a particular gene, chromosome, or part of the genome in an organism or cell with only a single chromosome complement. If you remember back to basic biology we were always taught that humans had $2N=23$, which meant that we had 23 pairs of chromosomes. Haploid cells and organisms are only $1N$. In humans we have mitochondrial DNA which have their own $1N$ or haploid genome. Thus when we do work with the mitochondrial DNA genome we describe the genetic data as haplotypes.

To simplify, we can think of haplotypes as diagnostic forms of a particular gene that can be geographically distinct. In organisms like wolverine there are some haplotypes that are only found in some mountain ranges and there are others that are broadly distributed. In southern Alaska and the Kenai Peninsula, Haplotype C is relatively common. It is not seen anywhere else surveyed so far. (One must be careful calling this Haplotype C, because names of haplotypes are not standardized. I know of 2 wolverine papers that have haplotype C's. The C we are discussing is in the Tomasik and Cook paper. Basically it is at the liberty of the author to call their haplotypes whatever they want [As, Bs, Cs, Ds, haplotype Green, Red, Yellow, etc.]).

Given that this haplotype is found only in Alaska and has never been seen east of there, and given that we know of Southern Alaska animals that were in captivity at one point in time in Minnesota, it seems likely that this is an escapee or released animal. On the other hand, until we sample Ontario and surrounds well, we can't be 100% sure.

The DNA results immediately brought one question to Jeff's mind: how in the world did this wolverine get to the Thumb from the Alaskan region, thousands of miles away? Audrey Magoun shed some light on it in an e-mail to Jeff dated April 3, 2006.

Here's my thoughts on what we know so far and what we should do: the cytC located in the Michigan wolverine indicates that it most probably originated from an area close to SE Alaska, but probably not SE Alaska because I don't know of anyone who has live-trapped wolverines there for zoos.

However, since they did not sample British Columbia or the Yukon or anywhere east of there, we should really try to have some material from Ontario tested too to be sure.

I really feel that the wolverine's genetic origins are BC or southern Yukon, which would correspond to the origin of most wolverines in captivity that I know about. However, I wouldn't want to put any % on my confidence level—just that from what we know at this point, the wolverine was probably a captive animal who either originated in BC or Yukon or had parentage that did.

I am going to try to get Ontario DNA sent to them shortly but I don't know how long it will take to run it—hopefully they will be interested enough to run it right away. On the sex, I hope we can get an answer to you today.

Jeff could almost taste the impatience like a sour lemon in his mouth. On Tuesday, April 4, he fired off a hasty e-mail to Tom Campbell, the editor of *Woods-N-Water News*. He wrote two versions of the story, one with the gender included and one without. What was the absolute latest he could turn in the article that Friday to meet the May issue's deadline?

Campbell was thrilled. He changed the May cover to get it on the front with a small photo attached. The caption read, "Wolverine's DNA. It's from Alaska? So, how did it get here?"

A local television station was pressing Jeff for the story, too, urging him to take a film crew out to the research site to tape a segment. But by this time, Jeff was fiercely loyal to the *WNW* team; he wasn't about to let anyone else break the story before it ran in Campbell's publication. Without batting an eye, he told the TV crew he'd be happy to oblige—but in order to protect the security of the site, they'd have to go in blindfolded. And, oh yeah, better wear rubber hip boots or waders just in case they broke through the mossy mat floating on top of the bog. They changed their minds.

When that issue of *Woods-N-Water News* hit the newsstands, no new father could have been any prouder than if he'd just given birth himself to a 40-pound ball of teeth, claws, and fur: "*It's a girl!*"

April, 2006

✦ ✦ ✦

Even though the initial DNA results seemed to conclude the wolverine was of Alaskan origin—and therefore most probably an escaped pet or zoo exhibit—Jeff lost none of his enthusiasm for studying her. In fact, he reached out to the public through the press and his website for any clues readers might be able to provide to solve that most basic mystery: how had a wolverine ended up in Michigan's Thumb?

"Let's find out where this gal came from," he wrote in a tongue-in-cheek plea in the May 2006 issue of *Woods-N-Water News.*

> Anyone driving through Michigan who has seen a large hole in a barn with visible claw marks, give me a call. Maybe she was a pet. Or if you find a large box secured to a downed parachute that says, "Gulo, do not pet" let me know. Maybe you've visited a zoo recently and found a hole chewed through the metal cage, with a sign attached saying "Wolverine—A solitary animal that is pound-per-pound the strongest, rarest, and most elusive mammal roaming the earth."
>
> Regardless of the origin of the Thumb wolverine, we've been blessed to have such an interesting and fascinating animal inhabiting the Thumb area over the last two years, and hopefully, if we all do our part, she will continue to stay alive and well.

He used the same article to put out a plea for financial help. In an effort to capitalize on their work, he and Jason Rosser had invested their personal funds in producing DVDs from their videotaped footage and were now $4,700 in the red. Steve Noble had also put together a PowerPoint presentation they were using for public appearances at high schools and sportsmen's clubs across mid-Michigan. The wolverine was no longer just some interesting wild animal they were studying; it had become a marketable commodity.

That might have seemed self-serving and mercenary if it weren't for one obvious fact: the only thing Jeff was spending the money on was more equipment and supplies for what he'd come to call simply his "research"—and what was coming in wasn't even coming close to matching the steady flow going out.

Meanwhile, the DNA support for the escapee hypothesis had captured everyone's imagination. In conversation with Audrey Magoun, DNR wildlife biologist Arnie Karr had raised the possibility that the wolverine had escaped or been intentionally released by someone who had acquired her from the rumored underground network of people dealing in exotic species without permits.

Jeff was extremely skeptical of that theory. He and Jason had made some attempts to locate such a network on the pretense of buying a wolverine, but they'd had no luck at all. If such people existed in Michigan, he had no idea how to find them.

While a wolverine escaping from black market thugs might sound like the plotline for a movie thriller, the notion isn't as far-fetched as it might seem. In fact, a highly publicized case made headlines in mid-Michigan barely a year later.

In November of 2007, the *Flint Journal* reported that three Genesee County people had been arrested and would be tried for stealing exotic animals from zoos, animal sanctuaries, and pet shops in Wayne and Saginaw counties, as well as other sites all over the country.[1]

Over a period of time, the Genesee County trio had allegedly stolen nearly 70 exotic birds, fish, and mammals, including a kinkajou, several toucans, a fennec fox, piranha, a rare two-toed sloth, and a wallaby. The amassed animals were confined in cages and enclosures inside a small Flint Township home for use in magic shows and school education programs. Bizarrely, the thieves even had a website advertising themselves as "Those Animal Guys"—which in fact turned out to be the key that eventually led to their arrest. The trio had apparently even set up a breeding ground for stolen toucans inside a hollow tree stump. A single toucan egg could have netted them 10,000 dollars on the exotic animal black market according to one expert quoted by the *Flint Journal*.

In November of 2008, Julie Lock of Flint Township was sentenced to three months in jail and five years' probation for her involvement in break-ins and theft from two Wayne County pet shops. She was also sentenced to five years' probation in Marion County, Florida, for her involvement in the theft of six toucans.

Her husband, Adam Lock, was sentenced to 11 months in prison for stealing 10,000 dollars' worth of flea and tick medication from a Flint Township pet store and at the time of his sentencing still had pending cases for animal theft in Wayne and Saginaw counties and Marion County, Florida.

Joshua Roberts of Mundy Township entered into a plea agreement and was sentenced to five years' probation and ordered to pay restitution.[2]

Of course, no wolverines were discovered in the trio's warehouse of stolen exotic animals, and no facilities had reported the loss or theft of any wolverines around the time of the ring's discovery, all the way back to 2004 when the Thumb wolverine first appeared—or even as far back as 2003, when the trapper Dale Abbott claimed to have first spotted wolverine tracks in the Minden City State Game Area. Still, the highly publicized case illustrates the fact that activities such as Karr described were indeed occurring in Michigan and might well be more common and far-reaching than anyone suspected.

Regardless, Audrey and Jeff agreed it would be wise to drop any further investigation into the animal's possible human source for fear that some wily opportunist might be inspired to take advantage of the situation and try to claim it as his or her own.

Bringing in a potential mate to try to start a breeding population also wasn't a good idea, said Magoun. As large and wild as it might seem to humans, the Minden habitat was simply too small to support any offspring—and their subsequent dispersal into the surrounding area would likely result in negative attitudes if farmers' chicken coops were raided and trappers found their traplines plundered by hungry young wolverines foraging for enough sustenance to survive.

Jeff agreed, but still he was convinced that the Thumb wolverine's continued survival made a good case for some kind of future introduction effort somewhere in Michigan, as he discussed in an April 5, 2006, e-mail to Audrey.

I agree that if there was an introduction that it should be done in a larger forested area, such as the Pigeon River State Forest mentioned earlier, or the Upper Peninsula where it is still quite wild. After thinking out your response, you are right, trying to start a new population in such a confined area is too risky.

The longer she is able to stay here and survive, the better argument we would have for re-introduction in other forested areas of the Midwest . . . Even if she died tomorrow, February 2003 to now is over 3 years, so she's a survivor. If she could survive 3–4 more years here, it certainly would provide ample arsenal and an example to refer to regarding wolverines not being killers of humans and cattle, and their ability to adapt.

There is the question of age though. Assuming she was mature in 2003, she would have to be at least 5 years old, and possibly older, although her situation with our supplemental feeding probably will prolong her life, especially during the meager months of winter when she is receiving a constant flow of food.

Audrey reassured him that he probably didn't need to worry yet about his wolverine dying of old age. "Five years is still young for a wolverine. They don't usually reproduce until they are 3 years old or so, and have been known to reproduce at 12 years in the wild and even older in captivity," she wrote on April 6. "Some have managed to live in captivity for as much as 20 years or so. You may be collecting road-killed deer for quite a long time!!!"

By this time, Jeff was gaining a reputation around mid-Michigan as something of a self-styled wolverine "expert"—but there were undoubtedly times when a conscientious scientist like Audrey Magoun had to wince a little at his penchant for colorful storytelling. After receiving copies of the two videotapes Jeff had used to create the commercial DVDs, she gently chided him for playing fast and loose with wolverine "facts." But at the same time, she didn't hesitate to give credit where credit was due, noting with interest that the amateur researcher might indeed have recorded what appeared to be a new aspect of wolverine behavior,

I was able to hook up a VCR yesterday and watch the two VHS tapes—very interesting. You mention Minden and where you found the tracks, so I was wondering if you have shown that video to others because wouldn't that mean that other people know where the wolverine is now. Or is that just a tape you made that isn't distributed anywhere?

Also, you mentioned in the tape that you have seen films of wolverines driving off grizzly bears, and attacking wolves, and that wolverines with their low center of gravity can get under a wolf and rip its belly open with its claws. I was wondering what films did you see and where did you get information about ripping bellies open? Most films of predator interactions with wolverines have been filmed with captive animals—I have only seen one clip that looked credible of a wild wolverine at a carcass with two wolves nearby that appeared to give the wolverine wide berth—it was a long time ago and was filmed in Canada or Scandinavia and was poor footage from quite a distance. I have never seen any evidence that wolverines use their claws to seriously scratch anything, including each other. In fact, I was astonished that the kits I raised never scratched me like domestic cats often do when you play with them. Wolverine claws are not particularly sharp—I'd say they seem halfway between cats and bears. They are used mainly for clinging, grasping, and digging (mainly through snow—not so much in dirt although they will make shallow small holes in dirt or moss sometimes to defecate) and I have never seen a large dirt hole made by a wolverine (compared to foxes and wolves). I was wondering where the idea that wolverines use their claws for disembowelment or digging large dirt holes comes from? I have heard other people mention that wolverine claws are long and sharp and dangerous, but I just don't see it in the animals I have been around—they definitely don't use them the way cats do. Just thought I'd mention it so that we can get the facts straight for the public.

Finally, I noticed in the night vision video several times when the wolverine was on or near the carcass, raised its tail, and then flicked it before lowering it. This is something they do when they defecate and yet you said you have never found their scats. Now I'm wondering if they also do that while depositing anal scent—I've never seen them mark carcasses with this scent and this would be new information if we can prove they are doing it because for now, we are assuming that is something that happens when they are disturbed or excited, not as a routine behavior around carcasses. If you ever see little brown colored droplets around the carcass, let me know or collect some of it in a little plastic vial and keep it frozen. We can try to find out if it is the anal musk.

Jeff wasn't offended in the least by her impromptu lectures. To the contrary, he felt lucky to be getting a free education from one of the world's foremost wolverine experts. He immediately responded by removing from his website any misinformation he'd posted about a wolverine's ability to take on a grizzly bear or wolf.

The novel behavior Audrey was referring to was some of the newest footage they'd captured. Back in March they had installed their first night vision video unit, and it was paying off with some privileged glimpses into the wolverine's nocturnal behavior. They watched, fascinated, as the wolverine moved about like a glowing-eyed phantom in the eerie green illumination of the infrared images.

The apparent scent-marking of a carcass proved to be a novel behavior that Jeff witnessed many times in subsequent tapings. He knew there was a healthy population of coyotes in the swamp—he often heard the packs yipping and howling when he went into the woods as dusk was settling in. But in all the time he'd been monitoring the wolverine, he'd never witnessed one come in to the bait site despite the lure of raw meat. If the wolverine was scent-marking the food, as Jeff suspected, it was obviously working.

Later that same month they'd also caught her on tape working to dislodge a carcass frozen to the ground, working her claws so furiously that ice and snow flew up behind her like a furry snowblower at work. Unable to budge it, she'd rolled upside down into the space she'd hollowed out beneath it, pushing at it from the underside, trying to break it free. When that failed to work, she gave up and simply gnawed at the frozen flesh. He marveled at the energy and ingenuity she showed, the resourcefulness that belied a keen natural intelligence.

Each new observation was serving to broaden Jeff's understanding and knowledge about a species that had been virtually unknown to him just two years before. He was gaining firsthand insights that many researchers spent a career obtaining.

Prompted by Jeff's questions after watching the video, Audrey explained to him what was actually known about wolverine digging behavior.

Wolverines make caches of food but they don't dig big holes in dirt—most caches are "made" either down in snow tunnels either at dens or snow tunnels similar to dens, in rock boulder screen, or in shallow small holes in moss or snow if the item isn't very large. They are also known to cache food in bogs in Scandinavia. Just putting the carcass into the moss or under water . . . When we say they "make" caches, we usually mean they carry food items to places that are secure from most other scavengers—into tunnels under rocks, downed trees, snow dens, etc. At large carcasses such as moose they may just tunnel down into the snow either under or near the carcass and stay there for a while or carry parts away to cache sites further away. The image that wolverines will dig a big hole somewhere where they then dump in a bunch of carcasses is not really correct. They may take quite a bit to a den site but the den site was probably the main reason for the cache being made there. A denning female is likely to take much of the food she finds to the den whereas a male wolverine is likely to distribute parts of a carcass around to different caching spots, but caching by wolverines has not been studied all that well. Most information comes from natal den (snow tunnel) sites or large carcass sites.

Meanwhile, they hadn't lost sight of their desire to compare the wolverine's DNA samples to another lab's batch of Ontario DNA to rule out any possibility that she had come from a region far east of the Yukon—a finding that would resurrect the possibility that she had migrated to Michigan naturally from the boreal forests in the north.

"Interestingly, they have a different take on the results that the first lab gave us," Audrey wrote in an e-mail to Jeff on April 8. "According to the second lab, that cytC marker is not going to be a good diagnostic of where the animal came from; so they think we should run it with the Ontario animals."

Despite some of the great footage they'd shot, Jeff and Jason mostly found themselves working their tails off for a scant few moments of tape. The wolverine was simply too good at removing the bait, grabbing it and leaving before the cameras had a chance to record much more than her furry behind.

Initially they'd just been setting a deer carcass in front of the camera then watching as she promptly hauled it away. They were going to have

The venison not only draws the wolverine to the research site but also encourages the animal to remain in the safety of the swamp, lessening the chance that it will be hit by a car, be poached, or enter farmyards and kill livestock.

to make her work a whole lot harder for her meals if they wanted to feature her in a starring role.

Jeff started by wrapping rope around the carcass, strapping it to a huge log lying on the ground in front of the camera. The wolverine simply used the log for leverage to pull the meat loose. She couldn't weigh more than 25 pounds herself. Jeff was in awe of the small animal's sheer brute strength.

Next he roped the carcass to a large tamarack tree, winding it around several times and snugging it tight with a knot.

The next time in, he'd find the meat gone, the stout rope still wrapped around the tree, its loops loosened where the bait should have been.

Later, when he watched the tape, he saw the fierce tug-of-war she'd won to haul the carcass off.

Next he and Steve brought a ratchet strap, wedging the carcass between two tamarack trees. While Steve pulled back on the smaller tree, Jeff wedged the carcass in and ratcheted it tight. The men could barely budge it between the two of them. They were sure they'd beaten her this time.

The camera told a different tale. That night she strolled in, hesitated for about five seconds, and seemed to consider her options in defeating the obstacle they'd set. Then she climbed the larger tamarack and came down toward the carcass headfirst, tugging it upward as her claws kept her anchored on the tree trunk. She had to work at it, but she managed to extract the carcass before the night was done.

After that, Jeff went to two ratchet straps, one high and one low. Finally he'd won, and she was forced to dine onsite. It was the technique they used from that point forward.

The bait issue had itself become a sensitive one. In the fall of 2004 Jeff had killed two bucks and a doe, so he'd still had a freezer full of venison to use as bait during the first year. But between the weekly baiting and his family eating venison four times a week, it was just about gone by the end of 2005.

He'd gotten the idea of supplementing the game meat with chicken earlier that summer, when he and Amy took the kids to a small zoo outside Frankenmuth to see its live wolverine exhibit.

They'd watched with keen interest as the zookeeper used whole chickens to feed the pair of wolverines. But Jeff soon realized that the daily feeding was pretty much the only time the wolverines were visible to visitors. The rest of the day they huddled in their shelter, hiding away from the humans' prying eyes. It didn't seem like much of a life to Jeff, who had seen firsthand the way a wolverine operates in a natural environment. It made him sad, seeing their limited existence and loss of dignity compared to "his" wolverine. The rebellious little boy inside him fantasized about returning at night to cut a hole in the fence, leaving a big arrowed sign pointing northeast with the words "Minden Swamp This Way."

He didn't do it, of course. He wasn't crazy. He was a respected teacher and responsible father of two, not some wild-eyed animal rights activist spraying paint on fur coats, unlocking tiger cages, and boycotting the circus.

But maybe he understood—just a little—why they did.

Discovering chicken as an alternative wasn't the solution he'd hoped for. It wasn't long before the DNR came calling to warn him to stop feeding the wolverine chicken. He was okay with that—he understood the reasoning—but the timing was bad. His store of venison was practically gone, and he wouldn't be able to replenish it until the next hunting season arrived.

He was driving to school when he saw an easy answer lying on the roadside in front of him: a deer freshly killed by a car. Jeff knew it was illegal in Michigan to be in possession of roadkill without a permit, so as soon as school was dismissed he contacted a local police officer and explained the situation. Since the permits were technically intended only for the actual motorist who struck the deer and wanted to take the meat home for food, the officer was willing to write him just the one permit.

Grateful for the small favor, he hastily loaded the deer in his truck and headed home. Using a truck winch to hoist the doe up in a tree in the backyard, he quickly gutted, skinned, and quartered the carcass so he could fit the sections in the big freezer in the family garage. Each 40-pound hindquarter would fit nicely in the metal-framed pack he used for hauling meat.

But he knew the dead doe would only last him a few months at best. He needed a more permanent solution.

The Caro police refused a request on the grounds that their permits only applied to a single incident of a car-deer collision. Next he approached the DNR. Arnie Karr, who had asked him to stop using chicken, was sympathetic but firm: Jeff would have to go the head of the DNR's wildlife division if he wanted any kind of permanent permit. Hopeful, Jeff drafted a letter seeking authorization from the state to pick up roadkill for wolverine research. Surely that would be seen as suitable grounds for granting him the necessary permission!

The letter came just a few weeks later. As soon as he saw the DNR logo on the envelope, he ripped it open, unable to contain his excitement. Now he would be able to legally access all the free food he needed to continue feeding his "pretty gal."

The typed letter was short and to the point, and Jeff was stunned by the words. His research was a personal endeavor, it said, not affiliated with the state or any authorized scientific research study. Therefore the request for a roadkill permit to feed the wolverine was denied.

Furious with disbelief, Jeff tore up the letter and threw it in the trash. He didn't mind that the state had never offered him any funding or physical help in studying the state's only known wolverine living in the wild—he knew just how underfunded and understaffed the DNR was. The wolverine might be an interesting anomaly, but it could hardly rate very high on the priority list for precious wildlife management dollars.

But he'd never expected it to toss a wrench so completely into his plans.

He paced back and forth, muttering and swearing as he downed first one beer then another and another in an effort to calm his rage. But with each bottle cap he twisted off, his angry resolve only grew.

He was utterly convinced now that the state wanted him to stop what he was doing. The longer he documented the wolverine's successful survival, he reasoned, the more credible would be his theory that this was truly a wild disperser, not some escaped exotic pet as the state maintained. Wildlife management was an expensive proposition. State wildlife officials weren't mandated to manage escaped pets; denying its natural existence was a pocketbook issue, he reasoned, nothing more.

It was a familiar line of logic, one that had been used by the "native cougar" proponents that Jeff himself disdained.

But at that particular place and moment, nothing else made sense to him.

He slammed down another beer, enraged. Nobody was going to stop him. Not the police, not the state, not *anyone.*

But he wasn't drunk enough to be stupid. He knew the only way he could continue was to find a legal solution. He wasn't about to give anyone an excuse to shut him down.

Bow season started in five months. Until then, he'd keep stashing road kill in the freezer, with or without a permit.

A single hunter could buy a combination deer tag for 30 dollars, allowing him or her to arrow two bucks. He could purchase tags for up to three does at 10 dollars apiece. If he took five deer during hunting season, he would have enough venison to continue his research for much of the following year.

Suddenly his bow-hunting rituals took on a heightened meaning, a whole new context. Not only would he be hunting to feed his family; he'd be feeding the wolverine, his "pretty gal," for another year.

The hobby he'd always approached like a business became even more intense. All the precautions he normally took to avoid alerting the deer to his presence took on a heightened, almost ritualistic importance: scent-washing his gear and clothes and storing them in airtight totes, showering before hunts with scent-killing soap and shampoo, spraying his outerwear with scent eliminator spray, wearing Scent-Lok bibs and coat, and shaving off all body hair on his armpits, chest, and groin that might capture and hold his human sweat and smell.

That fall he hung 31 tree stands throughout Sanilac and Tuscola counties, including three in the Minden City State Game Area. Every one of them would be crucial now. No hunter ever set out with more determination and sense of purpose.

He had to arrow five deer—and two of them needed to be mature bucks.

And that's exactly what he did for the next three years.

Summer, 2006

✦ ✦ ✦

In May 2006, the Human Genome Project announced completion of the final chromosome sequence in its 17-year initiative to map the genes of human DNA, marking it as one of the largest single investigative endeavors in modern science in terms of time, success, and scope.

It was the summer in which the solar system "lost" its ninth planet when the International Astronomical Union demoted Pluto to the status of dwarf planet.

It was also the year in which North Korea conducted its first nuclear test.

But out in the swamp of the Minden City State Game Area, it was none of those things.

It was, quite simply, the summer of the Raccoon War.

That spring Jeff and his crew had finally managed to solve the frustrating problem of the wolverine making off too quickly with their bait by double strapping the chunks of carcass several feet off the ground between two trees. The innovation was paying off with extended footage of her efforts to solve the novel challenges. One video showed her tugging hard on a 10 × 6 inch hunk of deer rib cage, diligently fighting to free it from its perch four feet off the ground.

Unfortunately, it also made it more difficult for her to protect her food source from other scavengers.

For quite some time, the men had suspected the wolverine was removing the meat from the bait site to cache it elsewhere in the swamp. Elsewhere, wolverines had been observed using rocky crevices and large deadfalls of trees to store their food caches, but those would have been easy targets for industrious raccoons. Audrey's best guess was that she was hiding the meat under water in the bog itself, where the odor would be masked. It was a technique researchers had observed in Scandinavian wolverines and made logical sense in Minden, with its similar boggy habitat.

But now the bait was staying in situ for longer periods of time, and the raccoons were having a heyday. The trail cameras became the *Raccoon News*, chronicling their antics. They were like piranha all over the research site, sometimes removing all the bait before the wolverine even got there. Jeff was getting tired of hauling in heavy loads of food only to have the raccoons steal it away. He even found himself getting annoyed with the wolverine. Why wasn't she taking care of the situation herself?

He tried to puzzle out the tantalizing bits and pieces recorded by the cameras. One segment appeared to show a raccoon running the wolverine off, then the wolverine returning two minutes later with the raccoon nowhere in sight. Three minutes after that, the tape showed the wolverine carrying the carcass away. The next recording was of a raccoon standing on its back legs, staring intently off in the direction in which the *Gulo* had disappeared.

It also seemed the wolverine had switched from nocturnal to daytime visits to the site—initially visiting in the mid- to late afternoon, usually from 1:00 to 4:00 p.m., but then showing up at all hours of the day and night.

What did it all mean?

Audrey's best guess was that since she could no longer easily remove the food to cache it in a more secure location, she'd taken to staying nearby to defend the site from the other scavengers. But they could find no evidence of daybeds: with the spring melt over, there was no snow in which to check for telltale hollows where she might be bedding down.

They decided to purchase a new voice-activated audio system, which might provide more clues into the interactions.

The payoff came one night in July, when suddenly the wolverine shot in a blur across the screen from right to left to come down hard directly on top of a raccoon. She brought the claws of her right paw down on the trapped raccoon, letting out a roar unlike anything Jeff had ever heard. He was roaring, too, as he watched it, cheering as if his team had just scored the winning touchdown. Finally she was letting the raccoons know who was boss in the Minden swamp!

Then everything changed almost overnight. At one point the bait

The wolverine wrestles a log weighing close to 100 pounds to gain access to bait buried three feet beneath it.

remained wedged between the trees for two and a half days without a single raccoon entering the site. It was only after the wolverine finally made off with the carcass that the raccoons returned.

It seemed the two sides had finally forged some wildlife version of a peace treaty. There was even one night when the cameras recorded the wolverine and a raccoon feeding side by side! Another daytime clip showed her running across the clearing with a raccoon following, almost as if they no longer took each other seriously as a threat—more like diners jostling each other at a buffet than hungry adversaries at war over scraps of food.

Jeff knew the wolverine had the size and strength to run the raccoons off if she wanted. Indeed, he'd recently witnessed a powerful testimony to her strength. As a test, he had placed a log 12 inches diameter and eight feet long over the bait, which he estimated at a weight of about 100 pounds. At first she tried to move the log with her paws, but when it rolled back onto the bait, she walked down to the far end, grabbed the end of the log with her mouth and lifted it up and over. Obviously, she was well fed and smart enough to choose her battles based purely on necessity. There was no question in his mind that if she was hungry and really saw the raccoons as a threat to her livelihood, she'd be dealing with them in an entirely different manner.

The Raccoon War, it seemed, was over.

But a new war was about to begin: a public controversy unintentionally launched by Jeff himself, which would linger long after the Thumb wolverine's death in 2010.

As a follow-up to those first genetics results, Audrey had thought it wise to have further samples tested against wolverine DNA from eastern Canada in order to provide a more comprehensive genetic picture of the Thumb wolverine. Jeff had eagerly supplied the samples, assuming the new results would verify the initial findings and confirm conclusively that she was of Alaskan stock. But the science of genetics is far too complex for simple answers in black and white.

On May 31, 2006, Audrey Magoun received the Thumb wolverine's DNA analysis from the second genetics lab, which added a new layer of complexity to the origins picture.

The new results were based on a program using 11 microsatellite loci—35 animals from Manitoba and 38 from Ontario, with the rest from the Northwest Territories and Nunavut, for a total of 220 samples.

The preliminary findings using that sampling base indicated that the Michigan wolverine was much more closely related to an Ontario or Manitoba animal than one from farther west in the Northwest Territories or Nunavut, Canada's Arctic region.

If those preliminary findings turned out to be correct, it could mean the Michigan wolverine was indeed a natural disperser that had moved into the United States from a region east of the Rockies.

The new message couldn't have hit Jeff harder than if he'd been physically punched in the solar plexus. Was it possible that the earlier findings were wrong and his "pretty gal" really was of relatively local origin?

If she had indeed come from Ontario, he reasoned, she must have come down into southern Ontario and then traveled westward into the

Michigan Thumb—or she'd headed south from Sault Ste. Marie, which wasn't that far from the Chapleau area where wolverine tracks had been spotted the past winter. That made more sense than thinking she'd traveled via Minnesota or Wisconsin or across a large expanse of Lake Superior or Huron.

Truth be told, it was hard to believe a female wolverine could have made it safely through any of those areas. Instead of a wildlife corridor, she'd have had to run a deadly gauntlet of agricultural, industrial, and suburban landscapes crisscrossed by heavily trafficked roadways And regardless of her route, she would have had to cross a Great Lake somewhere, which meant an ice crossing. It stretched credulity to the breaking point to think she'd strolled across on a toll bridge past US and Canadian border patrols.

Could more DNA testing help clarify the scenario of this mystery animal's origin? There had to be more they could learn from further DNA sampling before Audrey was willing to commit to any one definite scenario.

In a June 5, 2006, e-mail to Jeff, Audrey counseled patience until they learned more, explaining how tricky it can be to interpret data until one is certain that all the right factors have been considered.

> I think we should wait before releasing any more information about the origin of this animal until we either have something definitive or we hit a wall. The problem is this: mtDNA haplotypes provide information on long-term evolution so widely spaced populations should be different; microsatellite DNA is changing on a much more rapid scale and therefore gives information on recent dispersal episodes but also should be different between widely spaced areas.
>
> The second lab did not look at microsatellite DNA from Alaska, Yukon or British Columbia and therefore cannot say with certainty that such animals would not match up to Ontario/Manitoba wolverines; they're just assuming that AK, YK and BC would be closer to Northwest Territories, Nunavut, etc. And they didn't look at all the haplotypes either.
>
> The first lab didn't look at all haplotypes from all areas either and didn't look at microsatellites.
>
> So now we have disparate information that can only be sorted out by making more thorough analyses and even then, we could end

up with some unresolved questions. And even if we all agree that the Michigan animal is of Ontario/Manitoba origin, it still doesn't mean she absolutely dispersed from there because someone from Ontario could have sold an animal secretly to a facility in Michigan or surrounding areas from which she escaped.

I suggest we just let them work on this a little more and if you can get more hairs, it would probably be helpful in case anyone else wants to take on this problem. She should be shedding a lot right now but by July, it will be harder to pull hairs.

Jeff, on the other hand, could hardly contain his excitement. He was eager to publish the news, literally shout it from the rooftops if he could—but he also knew Audrey was wise in trying to rein in his enthusiasm. He'd already come out with one story announcing the wolverine was from southern Alaska. He'd look like a fool if he now said the complete opposite then had to recant again later. He was beginning to understand why scientists are so cautious about when and how they release their findings.

But by September, when they still had no clear answers, Jeff was done waiting. The banner headline ran across the front cover of the October issue of *Woods-N-Water News:* "New DNA Analysis Indicates She Could Be a Local Gal."

There was no mistaking where Jeff's personal opinions lay. The headline was only the beginning. After enthusiastically recounting his take on the second set of findings, he summed it up thus.

Holy crackers! This gal may have actually crossed the ice after all. Can you imagine that? For those folks that haven't been keeping up with current events, wolverine tracks were confirmed last winter by scientists studying wolves only 90 miles north of Michigan, in Chapleau, Ontario. In 2002, a GPS radio collared wolverine was documented traveling 258 miles, as a crow flies, in a period of 19 days, all while climbing up and down steep slopes in the mountains of Idaho. Huh! We may be on to something, but at this point any theories are only speculation at best.

Jeff had honestly hoped his article would provide an unbiased view of both sets of findings. But a poker face had never been one of Jeff

Ford's personality traits, and his writing was no exception. Nobody involved was pleased—least of all Audrey Magoun, who found herself stuck in an uncomfortable spot between her colleagues and this maverick amateur researcher who in one fell swoop had unwittingly cast an uncomplimentary shadow on the reputations of two highly respected genetics labs and managed to offend several important members of the scientific community involved in wolverine research.

It was exactly the kind of situation that scientists fear most when dealing with the popular media. The methodology of science is an exacting mistress, with rigorous care taken to confine any statements and conclusions within extremely specific parameters. It's why good science reporters are among the rarest and most valuable of journalists—writers who can paint in broad brushstrokes that reduce complex ideas into simple bites without blurring and distorting the subject.

Probably every scientist who has ever been quoted in the press has a horror story to tell of witnessing his or her words and work twisted beyond recognition by a casual turn of phrase or simple misinterpretation, like the childhood game of telephone. It's rarely malicious and almost always unintentional—but that doesn't make the resulting embarrassment any less real for the subject involved.

This wasn't really one of those times. But in his attempt to publicize the discrepancies between the two labs' findings in a way that lay readers could understand, Jeff had unintentionally oversimplified the significance and implications—throwing egg on both scientists' faces and unintentionally thrusting Audrey Magoun squarely in the middle.

Audrey wasted no time contacting Jeff, urging him to explain to his readers that when it came to wolverines genetic mapping was still a developing science and the seemingly disparate results didn't mean either lab was in error—only that more data were needed to correctly interpret and correlate the various results.

If anything, the second geneticist was even more dismayed than the first by Jeff's article. He told them he'd taken great pains to explain in detail that the two sets of results just provided different pieces of the same genetic puzzle. Any number of scenarios could explain it—maybe it was something as simple as an Alaskan female and eastern Canadian male having been bred in captivity and the resulting offspring released in Michigan. Who knew? Without another Michigan wolverine to test, or a larger sampling database that included wolverines from all the regions in question, all they could was speculate.

The second test had simply been unable to exclude the possibility that she was from central Canada, he said; that was quite different from saying the tests indicated she was from Canada rather than Alaska.

And Haplotype C, that unique genetic marker that had thus far never been found east of southern Alaska? The marker still hadn't shown up in about 20 samples from the Ontario control region used in the second set of tests, although some other very closely related northern haplotypes did. More tests could be run, using all the available samples from Russia, Alaska, Yukon, Idaho, and the various Canadian territories and provinces. But the results weren't likely to prove any more definitive given the level of random mating already observed in far northern populations.

Jeff was mortified to say the least. As a science teacher himself, he had the utmost regard for those in the larger scientific and academic community. The last thing he'd ever intended to do was embarrass anyone or cause a stir between respected peers. If anyone felt at all burned by Jeff's actions, none of them could possibly be feeling the sting of it more than he did himself.

All he wanted was to retreat into the world he knew, where everything made sense and he never had to second-guess himself or worry about what others said or did: the swamp, the woods, the rivers and streams that had provided comfort and companionship to a lonely, young, motherless boy so many long years before.

For the rest of that year he lost himself in the simple reverie of those familiar outdoor spaces, spending long hours hunting and fishing, alone or in the company of family and friends.

And of course, there was the wolverine.

In early 2007, deep in the heart of their second winter together, she gifted him with an unexpected glimpse into her secret and extraordinary world.

By this time there were all kinds of bones from deer carcasses littered around the research site. Rib cages stripped clean of meat. Leg bones and shoulder blades bleached grayish white from long days in the weather and sun. It was starting to look like a graveyard or ritual sacrifice site.

It was a night of deep, fresh snow. She came in to the bone-strewn site to remove some bait as was her usual routine. But something about a bare deer leg sticking up from the snow caught her eye.

She grabbed the leg bone in her mouth, rolled over on her back, and held the bone up between her front claws. Then she began tossing the bone into the air, catching it with her paws. Each time she dropped the bone she'd pick it up in her mouth and toss it in the air again.

Jeff watched the video over and over, amazed. There was no mistaking it. She was *playing*.

Watching the footage made him feel better than anything else he'd ever seen. He knew then that she was content with her life in the swamp, that even though she was alone and hadn't seen another of her kind for at least three, maybe even four years, that it was okay. She had everything she needed—or at least, she'd made the most of what life had dealt her.

Suddenly, all his childhood memories came flooding back, images of all the long hours and days he himself had spent alone in the woods, content and unafraid. He hadn't needed another human being at his side to feel happy. Mother Nature and all her treasures were enough.

So often he'd wondered if the wolverine was ever lonely for a family, for others of her kind. He knew wolverines were solitary animals, but studies had shown they often remain with their mother and siblings until old enough to find mates of their own, up to two years of age. Did she ever think of them? Was all she'd lost or left behind still somewhere deep inside her, a shapeless yearning emptiness?

But now here she was, playing with a deer's leg bone as if it were a child's toy. She was happy, living her life to the fullest. Content.

Jeff felt his bond to the wolverine deepen into something so intimate and real he could scarcely find words to describe it. They were the same.

2007

✦ ✦ ✦

For Jeff Ford, the third year began as cold and barren as the vast frozen farm fields of Michigan's Thumb. The digital cameras had logged a few pictures of the wolverine in the fall, but by October there was nothing. A single vague, faded track in the mud was the only sign he'd seen of her in endless weeks, and even then he couldn't be sure it wasn't just his own wishful thinking. It seemed to be the right size and have the correct number of toes, but even so, he couldn't be positive.

The lack of snow had persisted throughout November and December of 2006, making it impossible to determine her whereabouts as the ground grew ever colder and more unyielding to the passing of both man and beast.

Where was she? Was she still making a living somewhere in the frozen swamp? As each day passed without a trace, Jeff found himself again brooding over all the possibilities, none of them good. Was this the season she'd finally been dispatched by a startled hunter's gun? Drowned in a muddy bog, her leg caught fast in a beaver trap? Hit by a car as she crossed a dark stretch of road? As he drove the long, empty highways back and forth to school, he felt his eyes unwillingly drawn to the roadkill along the shoulder, compelled despite himself to seek out her familiar shape in every dark, furry mass.

Where was she?

He was at his wit's end. On January 22, 2007, he sent out a public plea on his website.

> I could use some help. Anyone living in Michigan who thinks they've come across a wolverine track please contact me and document the sighting with a picture and/or a cast to confirm the species . . . Wolverines can travel tremendous distances so the thumb

wolverine could show up anywhere, even out of state, especially during February and March while the lakes are iced over and travel is thus unimpeded . . .

. . . A wolverine's paw is large, 5 inches long by nearly 4 inches wide and the toe placement is laid out much like a human. They have what I would call a thumb off to one side, and a smaller pinky offset to the other. These are set on an angle, not even with each other, with the larger thumb set further back along the side of the paw. If you lay your hand down on a table with your fingers together, the basic layout of a human's hand is eerily similar to that of a wolverine.

A black bear or cougar track can easily be mistaken for a wolverine, but there are some identifiable differences. Although a black bear has 5 toes like a wolverine, and similar dimensions, especially as a young adult bear, all 5 toes are out front of the paw, whereas the wolverine has only 3 toes out front, and 2 offset on opposite sides.

The wolverine also has large claws that sink in off the end of each toe. These claws are often difficult to see in the snow, but easily identified in the mud. Identifying the toes on the side of the paw print on a wolverine walking in the snow can also be difficult, because they do not extend out very far, and often I've found even with a fresh wolverine track it's difficult to count all the toes occasionally, especially if in the deep snow.

One other identifying feature of a wolverine is its small gait between steps. Wolverines have short legs and their gait while walking is often 10 inches or less, while a black bear is much longer. The short distance between tracks and large print when laid out across an area immediately catches your attention as being unique or odd, and is a good indicator of a wolverine track.

Felines such as cougars and bobcats all have 4 toes, as do canines such a coyotes, wolves and domestic dogs. These species also have 2 toes evenly set out front, and the other 2 directly even with each other on the side, whereas the wolverine's thumb and pinky are offset at an angle just like the hand of a human.

All the characteristics of a wolverine track mentioned above have helped us to sort through the many tracks we find in the woods and along roadsides as we forage through the Thumb searching for signs of the wolverine. Keep your eyes open, and let us know if you come across a wolverine track. Hopefully, she is fine and healthy, and will continue to prosper here in Michigan.

The long dry spell finally ended deep in February, when the men again found her tracks in the late winter snow. Within two weeks, they were again getting pictures of her with two of the digital game cameras, and better action footage than ever.

Their new video system was equipped with an optional 24-hour surveillance mode that used a sensor to detect motion and heat, triggering the infrared lights without alerting the animals. The new system opened a whole new world to them, giving them a front row seat from which to view the wolverine's nocturnal habits, when she seemed most relaxed and casual, even carefree.

For Jeff, it was like watching his own private cable channel: *The Wolverine Nightly News.*

In one episode, the wolverine pounced on a raccoon and asserted her dominance before casually releasing it unharmed to scurry away. Another transaction made her dominance over the raccoons even clearer. Fascinated, he watched as a raccoon pulled some bait off the tree. Then the wolverine comes into view and slowly picks up the food and walks away with it as the raccoon sits off to the side, unmoving. It was the same kind of deliberate, cautious movement Jeff had seen countless times between two posturing dogs.

In one memorable solo scene, the wolverine jumps up in the air to grab a hunk of venison tied to a tree branch. A few seconds later, she is swinging across the screen like Tarzan, the branch held tightly in her mouth. At other times it seems she is simply playing in the snow.

Just as had been the case the last time she had vanished for a prolonged period, the team had no clue where she'd been, why she'd left or why she'd returned. She appeared healthy and robust, no worse the wear for her absence. Without the aid of a tracking collar, Jeff knew he had no way to solve the mystery. It made him realize how little he really knew about her behavior patterns, prompting him to begin keeping a careful record of her movements.

That year he analyzed data from 100 pictures taken randomly over the course of the year with his Trailwatcher DSC S-600 game camera, which had the most accurate internal clock.

The camera catches the wolverine leaping to reach bait secured overhead.

His goal was to discern any time patterns in her movements in and out of the research site, in hopes of better understanding the pressures and cycles that dictated the rhythms of her life. It may have lacked the rigor and precision of a true scientific study—but it was the kind of simple, observational record keeping that is at the heart of all scientific research and represents the only such effort involving the Thumb wolverine.

Wolverine Arrival Times at Research Site
# of times recorded per each time period

    12–2 a.m. - 2
    2–4 a.m. - 2
    4–6 a.m. - 6

6–8 a.m. - 17
8–10 a.m. - 16
10 a.m.–12 p.m. - 15
12–2 p.m. - 5
2–4 p.m. - 4
4–6 p.m. - 2
6–8 p.m. - 7
8–10 p.m. - 16
10 p.m.–12 a.m. - 8

Conclusion:

It was obvious to me that the Thumb wolverine was somewhat matutinal (42% of her movement was between the hours of 4–10 a.m.) although crepuscular may be a more accurate term because she was also fairly active in the evening hours around dark, with 20% of her movement between 6–10 p.m. Dawn and dusk movement accounted for 62% of her total movement. (62% crepuscular)

One fact I noticed during the six years of study was that during the hot months of summer (primarily July and August) she had a much stronger tendency to arrive at the research site late at night, or in the very early morning hours. These hours were when the temperatures were the lowest so I think this was one of her ways to combat the heat—by being most active when the temperatures were the most comfortable.

During the winter months she seemed to come in at any time, preferring daytime movement just as much or possibly more than nighttime movement.

Jeff's camera techniques were improving. too, with three game cameras and the video system functioning without a hitch, 24 hours a day. By this time he'd learned a new trick: burying the bait.

He'd set the camera down at wolverine eye level within one to three feet of the bait. He didn't get as many full-body shots now, but the ones he did get were fabulous, providing extreme close-up detail and a level of intimacy unmatched by anything he'd done to date. Still, in the winter months he was just as likely to return and find the lens buried in new snow. His solution was to set another camera a few feet higher up in the tamaracks to assure that one lens would remain unobstructed even after a storm.

That January Jeff also got a rare peek into her habits far removed from the prying eyes of the cameras.

He and Steve Noble were making their annual foray deep into the bog, well beyond the parameters of the research site, casually checking for buck tracks and signs of the wolverine.

They were walking along the west side of an island of tamaracks, about a half-mile south of the camera site, when they spotted something big, about 20 feet off the ground, wedged in the crook of one of the trees. When they got closer, they could see it was the front leg and shoulder of a white-tailed deer, hooves and fur intact.

Jeff didn't need video footage to know instantly who the culprit was: the wolverine had used her powerful jaws and claws to separate a hunk of carcass, then hauled the heavy cache up to where it would be safe from other scavengers. It had to have weighed 30 pounds! He couldn't stop grinning as he imagined the feat, even though it was one he couldn't document on film.

A few months later, though, he got all the proof he needed. He was watching a television program on wolverine researchers in Scandinavia. Sure enough, there it was: footage of a wolverine storing a wild animal in the crook of a tree.

Exactly like his "pretty gal."

By this time, Jeff really was becoming something of an amateur expert on the species he spent so much time studying, both directly and through the research work of others such as Audrey Magoun.

The bait setups had become almost an aptitude and agility test for the wolverine's problem-solving skills. They would hang the meat high up in the tree, forcing her to climb to retrieve it; bury it underground to see how easily she sniffed it out and dug it free; hid it beneath heavy logs to gauge her strength and persistence in freeing it.

Audrey had told him how much she'd come to admire the animals for their sheer perseverance, working very hard to accomplish whatever task they'd set their minds to. But the trait that had surprised her most was their playfulness. Even adult males displayed a wide range of personality types, and captive wolverines sometimes even formed unexpected bonds with unrelated wolverines that she could only describe as "friendships."

Jeff had witnessed that playful nature many times in the Thumb wolverine. How would she interact with another of her species? Was she lonely for the kind of camaraderie that Magoun and other researchers had witnessed even among this solitary breed?

Officials at the DNR had announced long ago that their intentions for the wolverine were simply to enforce the protection order and leave her alone to live out her natural life span in the Thumb. They had also made it clear that there were no plans to bring in a mate or try to establish a wolverine population. She was the first and only wolverine ever verified to be living in the wild in Michigan—and quite probably, it seemed, she would also be the last.

Audrey Magoun had told him that when there is no trapping or natural predators and food sources are good, wolverines can live into their early teens and even some as old as 12 years have been known to bear kits. Audrey herself had documented one female in the wild that was 15 years of age. Some in captivity have been known to live past 20.

By everyone's best guess, the Thumb wolverine was at least five years old. Jeff had no doubt that his "research" was providing her with a comfortable lifestyle that could well increase her life expectancy to something close to that seen under optimal conditions.

Fifteen years.

It was, Jeff thought, a very long time to be alone.

By spring, Jason Rosser had all but dropped out of active participation in the project. The heavy equipment operator had taken a new job in Detroit and didn't expect to be back in the Thumb until July at the earliest. Steve was still going out to the woods on occasion when he could spare the time from work and coaching duties and had put together a PowerPoint presentation and edited some of the videos for Jeff to use when he gave talks at clubs and schools. But for the most part, Jeff was now pretty much on his own.

No longer satisfied with seeing her on film, Jeff longed for another close encounter. That first year when she'd charged them out of the brush he'd been as startled as she was. He wanted a chance to really see her up close and in the flesh.

That July he began spending long hours sitting in a camouflage tent in the brush just out of sight of the clearing. Long years in deer blinds had taught him the kind of patience required for such a vigil. His bigger worry was whether she'd catch wind of his scent and simply avoid the site. Or, even worse, would she become tolerant of his presence as something harmless and thereby put herself at risk around humans in general?

But Audrey Magoun was cautiously optimistic.

It will probably take some time for her to get used to the idea that you are around—if you can't stay overnight, she may just patiently wait for you to leave before she comes in. . . . Wolverines can distinguish between individuals so the question would be—does she see you as less of a threat than other people who might be in the area? Wolverines in captivity will approach people they know, and avoid other people they don't know. However, they are all individuals in their degree of habituation to humans and it just isn't possible to know whether your presence would habituate her to just you or to people in general if at all. Sorry I can't be of more help.

But by late fall, Jeff had to admit it was hopeless. Once or twice he'd heard something stirring in the brush nearby but nothing more. The wolverine hadn't abandoned the area—that much he knew for certain. He was still getting great pictures—when he wasn't around. Too often he'd sneak in before daylight and patiently wait through her favorite visiting hours only to discover later that she'd arrived and taken the bait a half hour after he'd left.

Like a doting father, he studied and fretted over every detail in the images he collected. One October picture worried him. A prominent tuft of tan hair had appeared above her left eye, stretching over to the right. Was it her winter coat coming in? Or had she been injured? It might have been a tussle with a big raccoon. Maybe she'd even fallen from a tree. Who knew? They're tough critters, Audrey reassured him, and it was not unusual to see scars all over the faces of older wolverines.

His most recent picture of her was taken on November 14 at 9:34 p.m., the night before the opening of gun season for white-tailed deer. He'd baited her heavily early that morning, hoping she would take a long nap with a full stomach while the hunters were prowling about.

He headed back a week later to check on her. And was relieved to find evidence she'd enjoyed her meal. If she'd survived the guns of opening week, he knew her odds of surviving another year were good.

The winter looked to be a hard one. By the first week of December the snow was coming in thick, with another five inches piled on top of the five that had already arrived. For most creatures, the long Michigan winter is a hardship to be endured. For a wolverine so far south of its natural range, it had to be sheer heaven.

Early on the morning of December 26, while other men were sleeping off their holiday dinners or dutifully repairing already broken toys, Jeff was deep in the winter woods, backpacking in a Christmas present for the wolverine: half a deer carcass, including both hindquarters and part of the back. He'd shot the doe with his bow barely a week earlier, when a blizzard had canceled classes just before the holiday vacation was scheduled to begin. He skinned it for her and ratchet-strapped it tight to the tree, knowing she'd pull it down with far less effort than he'd used to tie it up. But he strapped it down low, at ground level, with the video camera over it. Who knew? Maybe she was a senior citizen by now. He didn't want her working too hard for her holiday meal.

The clock on the game camera showed she'd been there barely three hours before. All around him, fresh wolverine tracks covered the snow. It wouldn't be long, he knew, before another snowstorm blew in to bury those tracks in a fresh, heavy blanket of white.

Wolverine weather had arrived.

2008

✦ ✦ ✦

Some very odd events began to occur at the research site that winter. It all started with the carnivorous hares.

Jeff had always wondered why he got so many pictures of snowshoe hares at the bait site. It seemed odd to him that the rabbitlike herbivores would want to hang out in an area literally reeking with the scent of a predator.

In fact, a British Columbia study published in 1986 showed clear evidence that snowshoe hares will actively avoid feeding in an area where the scent of wolverine urine is present.[1]

In October 1980, researchers discovered that a wolverine had attacked hares caught in live traps during the last night of a monitoring session in a stand of young lodgepole pine seedlings near Prince George, British Columbia. Unable to remove the hares from the traps, the wolverine had simply consumed whatever body parts it could reach through the cage.

The gruesome scene had occurred only on the left half of the nine-hectare grid, so the intrigued researchers left the cages in place and monitored the ongoing results. During the first month following the attack, no hares entered traps on the left half of the grid while trapping continued to be successful in the control half. The avoidance behavior had declined by February, when comparable numbers of hares were again trapped on both halves of the grid.

Obviously, some odor had persisted at the site for eight weeks. But what was it? Had the wolverine marked the territory or the traps themselves as food caches, using anal gland secretions, urine, or feces? Or was it the presence of hare blood on the traps that caused their brethren to steer a wide berth?

A follow-up study in 1985 provided the answer. The controlled experiments indicated that wolverine fecal odor had no effect on hare behavior, while gland secretions had only a marginal impact. But wolverine urine odor proved to be highly effective in discouraging the hares from feeding on lodgepole pine seedlings, their favorite food in the spring and autumn.

Obviously, snowshoe hares were savvy enough to know a wolverine was a potential threat. So why were they willing to hang out in a spot routinely frequented by one? The answer was shocking.

In January, Jeff began getting pictures of snowshoe hares ripping meat off the bait carcasses. In some pictures the hares were so determined to tear off a chunk of meat that they were standing up high on their back legs to do so. It wasn't just a fluke incident: in that one winter, Jeff accumulated over 50 pictures of hares feeding on a deer carcass.

Now he realized they were actually attracted to the scent of the rotting meat.

What in the world was going on in the Minden swamp? The long-legged leporid was supposed to be a herbivore; any middle school science teacher worth his salt knew that. Yet here they were caught on camera like an armed robber in a 24-hour convenience store. Contrary to what every textbook had ever taught him, it appeared that under the right circumstances snowshoe hares could be opportunistic scavengers, omnivores that would consume dead, rotting flesh as eagerly as they would snip off a tender young branch or fresh green foliage.

Snowshoe hares should be serving as a mainstay of the wolverine's diet. Instead, the Minden swamp hares were fellow diners at Jeff's 24-hour buffet. If he didn't have it on film, most people would never have believed him.

About the same time he also started getting pictures of mysteriously ghostlike new visitor, so white it nearly vanished into the snowy background. On video it looked like a little piece of snow running around. It was only when the creature turned sideways that a pair of tiny dark eyes and the black tip of a tail could be seen. With its white body nearly invisible as it darted across the snow, the spots of black on each end almost appeared to be two tiny black mice chasing each other.

It wasn't until the game camera caught the critter in a still image and he was able to zoom in and study it that Jeff realized what else he had coming in to feast on the carrion. It was one of the wolverine's smallest cousins, the long-tailed weasel, decked out in its winter splendor.

Jeff could hardly believe his luck: it seemed his little dinner table in the swamp was now hosting both the largest and smallest members of the weasel family, and even a few carnivorous hares, all on the same day.

It reminded him again just how fortunate he was to be able to witness these scenes. One of the things that had always amazed him was the wolverine's skill at remaining unseen in an area with such a relatively high human population. Michigan's Thumb might seem rural and remote by suburban standards, but for a creature accustomed to the subarctic wilds of the boreal forest, it must have felt like a busy street corner on Manhattan's Upper West Side. He had spent literally hundreds of hours in this 8,725-acre region she called home—much of it spent actively looking for her—and he had only seen her in the flesh once for a fleeting second.

He was keenly aware that without the invention of new technologies such as motion-triggered digital cameras and infrared video units, all his documentation of her continued existence would have been impossible. That one winter day in 2004 would have been all anyone ever knew or saw of Michigan's lone wolverine.

Despite himself, he couldn't resist following where that line of reasoning led. If a wolverine could stay that well hidden for at least three years, what else might be lurking in Michigan's woods? Was she really the only one?

In February 2008, an Oregon State University graduate student named Katie Moriarty was conducting field research in California on martens, a small member of the weasel family. Like Jeff, Moriarty was using an array of motion-triggered game cameras set in wilderness sites throughout the Tahoe National Forest in the Sierra Nevada.[2]

She made the national news when one of her cameras snapped a photo of a wolverine, a species that had supposedly been gone from the state since the 1920s. It evoked many of the same questions that Michigan's wolverine had sparked. Was it a natural disperser that had traveled there from the nearest known resident populations 900 miles to the north in Washington or 600 miles to the northeast in Idaho's Sawtooth Range? Was it evidence of an ongoing population that had eluded detection because the technology and resources hadn't been available to detect it until now?

In the California wolverine's case, natural dispersion seemed a far more plausible explanation than it was in Michigan, as the mountainous

West has hundreds of miles of true wilderness still providing intact wildlife corridors.

Ever since he'd begun publishing accounts of the Thumb wolverine, Jeff had become the J. Allen Hynek of Michigan's weird wildlife phenomena. Hynek was an astronomer and professor who'd unintentionally become something of a pop culture icon for ufologists after serving as an advisor on several U.S. Air Force UFO studies from 1947–69, including the infamous Project Blue Book. Like Hynek culling through endless crackpot UFO accounts in search of some kernel of actual scientific proof, Jeff felt compelled to sort through endless phone calls, letters, and e-mails from people insisting they had seen a wolverine somewhere in Michigan, as unlikely or unreasonable as most of those claims seemed.

After studying their accounts, Jeff was usually able to convince these "eyewitnesses" that based on the size, coloration, and mechanics of movement they'd described, they more likely had seen a badger, raccoon, dog, or other common mammal.

But some accounts weren't so easily dismissed. Like Hynek's infamous swamp gas theory, sometimes the alternative explanations seemed more implausible than accepting the reports at face value.

Still, eyewitnesses are notoriously unreliable, regardless of whether they're reporting on wildlife, crime scenes, or UFOs. Unlike Jeff's trail cameras, the mind's eye often sees what it wants to see and the memories it records can be colored and edited far beyond any semblance of reality.

Jeff still believed the most probable explanation for Michigan's wolverine was that she had arrived alone via an ice bridge from northern Ontario. But he couldn't resist toying with the idea of a small native population that might have remained concealed in Michigan's northernmost woods due to the species' solitary nature. He used his website and freelance articles to solicit the help of other amateur naturalists, offering detailed advice on how they could distinguish a wolverine track from those of bears and other mammals, how to make a plaster cast, and even how to set up a bait site and install a camera trap.

Despite all that, no real evidence was ever produced of another wild wolverine anywhere in Michigan.

By this time, Jeff's dual obsessions with the wolverine and deer hunting were taking a tangible toll on his family life.

The Trailwatcher camera catches the wolverine doing what almost looks like a belly roll over a log in the research site.

Especially during the three months of hunting season, he was only getting three or four hours of sleep a night. From October through December, he was up at 5:00 a.m. so he could get to school an hour and a half early to prepare lessons for his chemistry, physics, physical science, and biology classes. He used his prep hour and lunchtime to grade papers so he would be free after school.

As soon as classes were dismissed, he'd take a quick scent shower at the school and then head directly to his tree stand with hunting bow in hand. By the time he packed up and left the woods at dusk, it was usually after 9:00 p.m. when he got home and the two kids were already in bed, fast asleep, tucked in by Amy. During hunting season, he wouldn't see Riley or Clint for days at a time.

In January his focus would return to the wolverine.

He always made sure he attended the early important family events, but he had to admit Amy was carrying the heaviest share of the parenting load, especially through the long winter months. The summers became the family's emotional oasis, when both he and Amy were freed from teaching and could spend real family time together despite the long hours Jeff continued to devote to the wolverine.

In many ways, though, the wolverine had become a member of the family, maybe not physically present but nearly as tangible as if she'd been set a regular place at the family table. As the kids grew older, their interest in Daddy's invisible friend also grew. Whenever Jeff brought home new video footage, they'd sit down to watch it as eagerly as other kids might watch a new Disney movie or Saturday morning cartoon.

Each week he'd print the new photo images, too, and often the kids would help organize them into one of the nine photo albums he'd amassed over the years. Their questions were an unending source of amusement and joy: "Daddy, does the wolverine go to visit your school? Why don't you bring your friend the wolverine home so I can meet her? How does the wolverine go poop if she lies out in the woods and has no diapers? Daddy, do you give the wolverine kisses and hugs like me and Momma?"

That winter, Jeff began to notice some changes that hinted the wolverine project was taking a mental and physical toll on him too. The nearly three-mile hike in and out of the swamp was starting to feel like a burden. But things had been going so smoothly for the past year that he found it easy to ignore the signs at first, dismissing them as the small, inevitable changes of getting older. Heck, he was 43 now. Why should he expect to feel like he did when he was 32? Maybe it was even just a mental thing; maybe, like the other guys, he was simply getting tired of the whole routine.

He started packing in less weight, stopping to take off his pack and rest three or four times as he made the slow walk in.

It was frustrating. He'd always stomped in at a good pace, even hauling a 40-pound pack, sinking in the mud and slogging through heavy snow without so much as a pause in his step. But now, even carrying a lighter load, he was walking much slower and still having to sit down and rest, leaning his back against the pack while he caught his breath.

Instead of being the highlight of his week, he began to dread the weekly hike into the swamp, trying to invent reasons to avoid the inevitable. At the same time he grew increasingly angry at himself for what he interpreted as a lack of motivation. He cussed himself out for being a wimp, for losing the stubborn focus and intensity of purpose that had always been a source of personal pride. Over and over, the words of his old high school football coach played in his head, telling him to "suck it up," just like when he was dead on his feet and hurting but it was the fourth quarter of the game and no time to quit.

By the end of the summer, it had worsened to the point that every trip to the swamp was followed by a day or two of complete and utter exhaustion, when he could barely summon up the energy and enthusiasm to do much more than lie sprawled on the couch, watching TV. Amy interpreted it as a loss of interest in the family and began to grow increasingly resentful. Who could blame her? There he was, traipsing out to the swamp every weekend but unwilling to do a thing with his wife or kids for days afterward. The tension was building both inside and around him, a growing sense of unease.

Although he wouldn't admit to anyone—could barely admit it to himself—there were days when he seriously considered quitting the wolverine entirely. Just pack up his cameras and be done with the whole endless, pointless project. Did any of it matter to anyone but him? He wasn't a real researcher, just some guy who loved nature and hunting. There was no university or government agency waiting for the results of his years of study. Maybe it was time to quit.

In truth, it was all a way to avoid a larger fear that was getting harder and harder to deny. Jeff was starting to worry that there really was something wrong with him besides simple lethargy. Beyond the fatigue, he was having unexplained bouts of chest pain and palpitations that left him breathless and weak at the slightest exertion. But he told no one, as if denial would make it unreal.

Then, in August 2008, he suddenly lost his voice. At first he figured it was a temporary thing and his voice would come back soon, just a weird case of laryngitis. But after two weeks he began to wonder what was really going on. What other kind of mysterious malaise could be plaguing him now?

On September 22, he woke up in the middle of the night barely able to breathe. Every shallow intake of air was little more than a wheezing gasp. Then he began coughing up thick green phlegm from deep in his lungs. He managed to call the school secretary to request a sub, and that morning he drove to the medical clinic in Caro.

The doctor told him he had the flu. But by now Jeff's state of denial had turned into a dark and suspicious dread. Two of his uncles had died of heart failure, one in his late forties and the other in his early fifties. Deep inside, he knew that whatever it was, it was something to do with his heart. He convinced the doctor to run an electrocardiogram. A few minutes later the doctor returned with a very serious look on her face and explained that the test showed a "left bundle block" and other electrical issues in the left ventricle of his heart. He needed to get to a hospital right away.

Later, at the Bay Regional Medical Center, Jeff and Amy stood side by side staring into Jeff's chest X-ray, which the doctors had placed alongside an X-ray of a normal heart for comparison. Jeff's was nearly twice the size.

The laryngitis was a telltale sign of his condition, a doctor explained. People with a severely enlarged heart sometimes lose their left vocal cord. It becomes paralyzed because the nerve leading to the cord is pinched, completely cut off due to the heart deformity. Jeff wouldn't be going home anytime soon.

On September 26, the hospital performed a heart catheterization and echocardiogram and found that Jeff's heart was functioning at less than 10 percent. That meant that for every cup of blood that entered his heart, only two tablespoons were leaving out the other side to carry oxygen and nutrients to his brain and muscle tissue. The surgeon told them he was lucky he wasn't dead—it was a wonder he hadn't suffered full cardiac arrest while he was alone out in the woods.

But why was a 43-year-old physically active man's heart in such dire shape? The doctors probed him for more details on his family history. He knew about the two uncles on his mother's side that had died fairly young. But there was one close relative about whom he knew almost nothing: his paternal grandfather, Irvin.

Nobody talked about Irvin, who had run out on Jeff's grandmother, Faye, when Jeff's dad Jac was three years old, leaving her with two young boys to raise. It turned out to be a favor: otherwise she'd never have met and married Charlie, the beloved family patriarch who had taught his stepsons and grandsons to hunt and fish.

Jeff had seen pictures at the family cabin of his Grandma Faye when she was young. As a young girl, she'd been a trapeze and high-wire performer in the Flying Melvoras, part of her family's traveling circus. But Jeff had never so much as seen a picture of her first husband, his wayward Grandpa Irvin.

Jeff's sister Teri offered to do some investigating. She quickly learned that when Jac was 18 he'd received a call from Irvin's widow in California. Irvin had keeled over and died at the age of 42. He'd had a rare form of cardiomyopathy, she said, and she felt compelled to tell Jac in case it was genetic.

And now here was Jeff at the same age, diagnosed with the same condition. Toss in 20-odd years of cigarettes and hard living, and here's what you got: Jeff Ford, lying in a hospital bed, nearly dead.

In Jeff's case, electrical signals were not reaching key areas of the heart muscle so that the lower left side of his heart was not beating in sync with the rest. As the inefficient heart struggles, it enlarges due to the stress of beating out of rhythm. The hospital immediately scheduled him for a pacemaker implant. He had the surgery on September 29.

By this time he'd been hospitalized for a week and was going stir-crazy, lying in a hospital bed with way too much time to think. Would he be around to see Riley graduate from high school? Would he even be around to see her get out of elementary school? What about Amy? How would she raise Riley and Clint on her own?

He thought back to his own childhood and the loss of his mother at such a young age. How many scars would he leave on his own friends and family if he died now?

And what would happen to the wolverine? Had she become so dependent on his offerings that she'd be unable to fend for herself? It was only during the winter that Jeff took whole deer out to the swamp. During the other three-quarters of the year, the raccoons often raided the food cache ahead of her, so she had to be filling out the rest of her diet with the other food in abundance in the swamp. He was pretty sure she'd make a comfortable living on her own. But even his own father had questioned the wisdom of feeding a wild animal. He'd always wondered how long she might live. Now what would happen if he died first? Had his own actions doomed her right along with him?

Because of its extreme size, the surgeons were unable to get the third lead wire implanted in the bottom of his heart for the pacemaker and defibrillator. They scheduled a second surgery a week later and sent Jeff home to recuperate and rest.

As Amy drove him home, he told her he was going up to hunting camp. It was the start of bow season, and Jeff had never missed an opening day. It was almost a religious custom for him. She knew she couldn't stop him, but she made him promise he wouldn't hunt. He dutifully agreed, saying all he wanted was to rest on the couch at camp and listen to the radio.

She didn't see him pack his bow. He knew he probably couldn't pull it back, but it didn't matter. He needed to get back out in the woods and think, and the bow was like a natural extension of him, a companion. Almost the way a Buddhist monk might spin a prayer wheel, just handling the bow would help clear his mind and sort out the jumble of worries in his head.

That night, Amy's sister Jen's husband Bob showed up at camp. Jeff suspected his wife had "hired" his brother-in-law to watch over the prodigal patient, but he didn't really mind. And Bob understood why Jeff really needed to be there. The next morning they climbed into Bob's truck in the dark before dawn and rode out to the shed down the south trail that served as his hunting blind. Like Amy, Bob made him promise not to hunt.

"You can't draw that bow, Jeff. You might move those wires in your heart."

Jeff just grinned and shrugged it off.

"I won't draw it unless it's a monster buck," he said, chuckling.

But the reality turned out to be that he spent nearly the entire opening morning of bow season dead asleep, too weak to stay awake. At 11:00 a.m. he went back to the cabin and slept the rest of the afternoon.

When he got up that evening he was sore and miserable and lonely. The reality of his situation suddenly hit. He was acting like a sulking, stubborn little boy—not a husband and father with adult responsibilities to face. He couldn't even stay awake in the woods. He shouldn't be at camp even pretending he was able to hunt. He should be home with his family.

He scrounged around for paper and started writing a letter to his daughter Riley. If he died, he wanted to leave something behind that would tell her how much she meant to him.

Riley,

I want you to know how much daddy loves you. You're getting to be
such a big girl, and you are so beautiful and full of life. It seems like
yesterday I was rocking you in the chair—you know—the one with
all the fishies on it. As I would rock you and sing you a song, you
would look up at me and smile with those beautiful blue eyes. And
now you are getting so big.

You are a special girl Riley Jo Ford, and I love you so much. I'm
coming home in the morning. See you then.

Dad

It was to be the first of many letters he'd write over the coming
months, little notes tucked in random places for her to find.

The next week Jeff was back in the hospital for the second surgery.

This time they went in through the ribs below his left pectoral muscle to implant the third wire and hook it into the transmitter. They also
found some problems with the tissue in his lungs, and did a lung biopsy,
implanting a two-foot tube in the bottom of his left lung to drain excess
blood and fluid. He was immobile for 10 days.

During Jeff's surgeries and recuperation, Steve Noble stepped in
unasked and took over the reins of the wolverine project. He drove to the
house and picked up the memory cards, batteries, backpack, bait, GPS,
keys to the locks—everything he'd need to continue Jeff's research.

Steve went out into the swamp twice during the period of surgeries
and recovery, tending to the cameras and replenishing the bait, and
called every day for weeks to check on his friend's condition. Over the
past year or so, Jeff had to admit he'd come to doubt Steve's devotion to
the wolverine, had maybe even resented him for it at times. Now he
was humbled to realize that he had no reason at all to doubt Steve's devotion to him.

Jac Ford was also a frequent visitor during Jeff's recovery, dropping
by the house to make lunch or dinner for the family and then slipping
quietly back home.

Jeff was grateful for all of it. But he felt inadequate too. For the first
time in his life he was truly helpless. He couldn't work, couldn't rough-
house with his kids or help around the house, couldn't hunt or fish or
anything he'd spent a lifetime learning to do.

Surprisingly, though, he began to see something positive coming out of the whole mess. Because he wasn't working and was forced to stay home and out of the woods, he was spending more quality time with his family than he had in years. The bond with Amy and the kids was growing stronger than ever before.

By the third of week of October, though, Jeff was starting to grow restless. He could get around now without much pain, and his thoughts kept drifting back to the wolverine. Steve was keeping things rolling, but he knew that was a temporary fix. It had been weeks since he'd visited the wolverine, and he was worried that the lack of regular baiting would compel her to wander out of the swamp in search of food. He set his sights on getting back to the swamp.

He didn't dare tell Amy or his dad. He knew if the hike didn't kill him, they'd do it with their bare hands. Steve tried to talk him out of it. He wouldn't listen.

He was still under doctor's orders not to walk more than 100 yards or lift anything heavier than 10 pounds. Yet here he was, pack strapped to his back with seven pounds of bait and batteries, heading to the swamp in knee-high boots to see his "pretty gal."

It was a slow, tough trip out and back, as slow and hard as that trek through the blizzard had been in March 2005. Except this time there was no snow to battle against—only the limitations of a sick, stubborn man who never let anything stop him, and wasn't smart enough to let it, even when he should.

As he trudged along, he chanted one of his favorite Iron Maiden songs, from the *Piece of Mind* album, "Die With your Boots On." Every now and then, he'd stop and lie down on his back with his head propped against his pack, gazing up at the sky, praying for the strength to make it.

As he made the turn at the one-mile mark, he started to enter the bog, his boots sinking in the muck above his ankles. Every step was a chore, each boot making a sucking noise as he pulled it free of the muck to take another step, over and over.

He was extracautious weaving his way around the holes where the sphagnum moss met water. He knew one wrong step and he'd be sunk in the bog up to his waist. Then even if he managed to climb back out, his wet, muddy pack and clothes would weigh 30 pounds more for the remainder of the trip. He knew he couldn't afford that kind of mistake.

As it was, he'd need every ounce of energy he could muster to make it back out before Amy and the kids returned from school and discovered he was gone.

When he finally arrived at the bait site, he lay on the bare ground, exhausted. He'd been lying there about 10 minutes, next to the big tamarack where he always strapped the carcasses, when he heard the scurrying of a small creature.

He looked up. It was the long-tailed weasel in its dark summer coat, staring down at him from a branch on the tamarack, barely six feet away.

As he shifted to get a better look, the startled critter sprinted down the tree and vanished in the thick brush.

If he lay there long enough, could he be lucky enough to see the wolverine? He amused himself imagining her coming in close to where he was lying, sniffing at him curiously and maybe licking his hand the way his cat Smudge did. Chuckling, he struggled back to his feet to check the cameras and get a look at the pictures shot during his long absence.

After scrolling through 30 or more pictures of raccoons, there she was—his furry friend, the pretty gal, in her splendor. She was stretching out her huge front claws, stepping up onto the log he'd been lying next to just moments before. She'd been that close.

An adrenalin rush shot through him. He was back in the woods with the wolverine, back where he belonged.

The trip back out was equally slow and tiring. He had to stop numerous times to rest. But once he saw his truck in view down the trail, his pace picked up and his spirits soared. He'd done it. The research would continue.

Everything was going to be okay.

Last Dance, 2009–10

✦ ✦ ✦

The spring thaw hit the swamp with the hard rains of April, making progress into the area thicker, wetter, and muddier with every step. Jeff scaled back his weekly scheduled visit to every two weeks, telling himself it would only be temporary until the spring rains had passed.

But with each passing weekend it became a little bit harder to shrug off the weird mix of fear and discomfort that overcame him as he prepared to go in. The surgeries and enforced convalescence had left him with an entirely new and foreign sense of his own vulnerability and mortality. He dreaded the physical challenge now, was afraid that somewhere along the trail his heart would give out. It was a relief every time he made it back to the edge of the woods and saw his truck there waiting for him.

Still, Jeff continued to acquire images of the wolverine throughout 2009 and worked on getting more hair samples for a possible third DNA analysis. But he had undeniably lost some of his stubborn enthusiasm for the wolverine project. Thanks to the pacemaker, some of his strength and stamina had returned, but he knew the surgeries hadn't cured him; they had only bought more time. His heart would continue to deteriorate and eventually he would need a transplant to survive. It gave him a growing sense that time was running out to accomplish something meaningful out of all this.

In September of 2009, he wrote this e-mail to Audrey Magoun.

Things are well here. I've continued to monitor her and she is still staying in the same area (the Minden bog) although I hear stories of people who run into her on occasion outside that 6,500 acres. She sure is an adapter, an animal that supposedly needs so much wild territory living freely next to humans for so long. At no time in the last 5 years has she been more than a couple of miles from a human, yet she continues for the most part to remain secluded, isolated and continues to make a living.

I'm hoping someday her example will be a solid argument for reintroduction of wolverines to areas that traditionally were considered not "wild" enough. She sure makes a strong argument.

My most recent picture was August 28th, a birthday present from her to me I suppose. She looks very healthy and you wouldn't believe the size of the hole I'm putting bait in now. In some of my pictures all you can see is her tail sticking up in the air like a *gulo* flag. I'll send some pics when I get some time. Teaching 4 different science classes this year, including 2 different chemistry classes so my spare time is limited.

Audrey still had no idea about Jeff's health problems—he'd chosen not to confide that to anyone outside the closest family and friends—but she sensed his growing need for reassurance that this "project" he'd chosen to devote his life to was indeed worthwhile. She always found the time to be supportive even though she was heavily involved in her own research work, shuttling between seasonal fieldwork projects in Alaska and Oregon.

I think your work with this wolverine will be very valuable; I only wish you had good records on just how much bait you took into her area and how often. I know there will be skeptics that will maintain she wouldn't have been able to survive without your addition of food to the area for her. I don't know how much longer you can expect her to stay there but possibly many years. Would it be too much trouble for you to weigh your baits and classify them by what the bait is and keep a record of the dates you took the bait into the area from now on? Then I think we could actually write up a paper on her after a couple more years of keeping track of her. It would be very valuable for the reasons you stated. I'm looking forward to seeing more photos and when I get my manual finished, there should be lots of photos to look at from my study area too.

Neither of them could know that time was indeed running out—but not in the way Jeff might have expected. On February 16, 2010, he sent this e-mail to Audrey Magoun.

Hiked in yesterday to check for hair. Three clips were triggered, with only one containing anything—a talon from a small bird. One set of *Gulo* tracks coming in to the mock cache from the northwest, and departing down the same run to the northwest. But the only pictures from two cameras overlooking the run pole indicated a raccoon, weasel, and some birds.

Tried to remove the bait from the food cache but it was so frozen in the ground that was impossible.

She will eventually climb the apparatus and hit it; I just have to be patient. All the bait I lay will be overlooking the hair snag until I am confident I have sufficient quantities for DNA analysis, although I have been concerned why she's been leaving bait the last two times in; especially considering the duration between one trip was a month.

All the bait I had overlooking the hair snag was still present, so I packed out the bait I brought in for next time in two weeks.

It is extremely rare for her to leave bait between trips in. She has done it three times in a row now. Any ideas why?

The last photographs of the wolverine alive were taken on February 16, 2010. Her body was found semisubmerged near a beaver dam on March 13, less than a mile from Jeff's cameras.

News spread quickly in the days following the discovery of her body. Once again, Jeff found himself at the middle of a local media feeding frenzy. Wildlife biologist Arnie Karr found himself back in the limelight, too, inundated with questions about the cause of death and plans for the body. Almost immediately, people began debating the location of any future wolverine display. Some argued the wolverine's body preserved with taxidermy rightfully belonged to Sanilac County and should be housed at the county courthouse. Others campaigned for a state park facility, while still others insisted it belonged at the state Capitol or on permanent traveling display.

Coincidentally, Jeff had just published a three-page spread in the March 2010 issue of *Woods-N-Water News*, which included a dramatic retelling of the close encounters he, Steve Noble, and Jason Rosser had had with the wolverine in 2005 along with his most recent anecdotes. It ended with this author's note.

Forty pounds of venison hindquarters, ratchet-strapped to a tamarack tree just before a big snowstorm, ensures that the wolverine will have plenty of meat when other food sources are scarce.

Watch out for a future update in *Woods-N-Water News* on new DNA evidence. In January a wolverine was confirmed on Manitoulin Island, Ontario, a mere 50 miles from Alpena, and only 122 miles from the tip of the Thumb of Michigan. The DNA combined with the close proximity of wolverines to Michigan in Chapleau and Manitoulin Island are pointing more and more to the Thumb wolverine being a traveler, a vagabond, from Ontario on an ice bridge. We are currently collecting hair for comparison and analysis to wolverines from those areas.

Unfortunately, Jeff's enthusiastic reporting of the Manitoulin events had gotten ahead of the facts. On February 5, after the story went to press, Audrey Magoun informed him that the Manitoulin sighting was no longer considered "confirmed" and was in fact under serious scrutiny.

The information I've gotten back so far on the Manitoulin wolverine photos (which isn't much) is beginning to look like a hoax on someone's part. But I haven't found the photographer to speak to him/her directly. If you look at the photos carefully, it appears the wolverine came around twice (at least) in front of the camera, which suggests a set up with a captive animal. Furthermore, a biologist in Thunder Bay says he thinks he's seen these photos a while ago. Everyone is very skeptical now and no one is going to believe it unless we find the photographer and he can explain the situation under which the photos were taken. I don't think anyone would be willing to put a lot of cameras out without some indication an animal is sticking around. All this was disappointing of course, but I still think wolverines are moving eastward and there may be wolverines at least occasionally that far south. At least we know we have one in Michigan!

That wasn't the only thorn the article had stuck in someone's side. Apparently Jeff wasn't content to simply fan the flames of the DNA controversy and continue to annoy geneticists; the article also didn't endear him to state wildlife officials by reissuing his earlier public call for a wildlife introduction program.

The Thumb wolverine continues to thrive and is making a living in a habitat surrounded by humans; a habitat previously thought of as not enough expanse of wilderness to support an animal such as a wolverine.

Now we know that a wolverine can co-exist with humans in a fairly populated area and thrive. She is never more than a few miles from humans at any given time, yet has been able to live among humans for at least the last six years in the Thumb of Michigan.

I propose the Michigan DNR should initiate the re-introduction of wolverines as a population into Michigan. Three years ago they stated they are adamantly opposed to providing the Thumb wolverine a mate, as well as any re-introduction efforts in the Lower or Upper Peninsula.

With the DNR's successful reintroduction of elk near the town of Wolverine in 1918, and the recent successful reintroduction of wolves into the Upper Peninsula, it's time to also re-establish a population of wolverines. Humans were responsible for the loss of the thriving populations of elk and wolves that used to call the woods of Michigan their home in the 1800s, and as a result the DNR fulfilled their responsibility to return those two species back into the Michigan ecosystem. The wolverine also used to thrive in Michigan, and therefore the DNR also has an obligation to return this species back into our Michigan ecosystem, therefore creating a more rich and diversified ecosystem.

Although the Thumb wolverine at times has occasionally had my nerve endings on edge and causes a few goose bumps, I wouldn't trade that feeling for anything. I feel proud to walk the woods where a top of the line predator lives, and enjoy being part of an ecosystem where I am not the dominant predator. Wouldn't it be neat for other folks living outside the Thumb to also be able to walk through their woods and have an opportunity to see a wolverine, the rarest mammal in North America?

As it turned out, however, none of it really mattered. Before Jeff's words had grown cold on the newsstands, a bigger story had eclipsed them: the Thumb wolverine was dead, leaving the mysteries of her origins forever unresolved.

Or were they?

If the wolverine's death had gone unrecorded and the body never recovered, that would indeed have been the end of the tale. Instead, it appeared that in death she had left researchers a rare gift: the physical keys with which to unlock her secrets.

The implications of that didn't escape Jeff—but in those first few weeks, he was too overcome with grief to care.

On March 13, 2010, he sent a one-sentence message to Audrey Magoun: "The pretty gal, my friend, the thumb wolverine, is dead!"

He followed up with this longer explanation the next day.

Thanks for your kind words. Never realized how much her death would impact me, knowing it was inevitable. Find myself breaking into tears at the mere thought of her being caught in the beaver dam. My daughter was an angel today (7 years old) giving me big hugs and kisses, sensing I was not in good condition.

Steve Noble and I got specific directions on the death site, and were able to find the beaver dam early this morning. The beaver dam is the second one coming off Palms Road where I enter from the east side of the swamp/bog. There is a ditch slightly set back in the brush that parallels the trip in west to the research site. I may have walked right by her Thursday, or it happened shortly after.

The conservation officer, Seth, feels she died within the last 3 days based on her good condition. She was lying 2 feet off the lower end of the beaver dam (downstream) in the water 2 feet from the dam. We saw no signs of beaver trapping there; it is quite a drop and seemed to be fairly undisturbed, although I am going to check out the local trappers.

I was able to check her over thoroughly and found nothing abnormal, other than one of her teeth broken off, and some mud and debris in her mouth.

The hunter that found her thinks she tried to enter the beaver dam from the lower section and became stuck.

On March 15, 2010, the Michigan DNR issued the following press release.

Michigan's only known wild wolverine has died.

A female wolverine, first spotted in the Thumb in Feb. 24, 2004, was found dead by hikers at the Minden Bog in the Minden City State Game Area Saturday.

Todd Rann of Marysville and Morgan Graham of New Baltimore spotted what they thought was a dead beaver, partially submerged in the water near a beaver dam. Rann pulled it from the water and realized it was a wolverine. The pair called the Report All Poaching hot line. Department of Natural Resources Conservation Officers Seth Rhodea and Bob Hobkirk responded immediately and retrieved the animal.

The officers reported no visible signs of trauma.

DNR wildlife biologist Arnie Karr, who originally verified the animal was a wolverine after it was treed by coyote hunters in 2004, said the carcass will be sent to the DNR veterinary lab for necropsy. The department plans to have the specimen mounted and displayed, probably at the visitor center at nearby Bay City State Recreation Area, Karr said.

The animal was the first wolverine ever actually verified in

Michigan. Biologists say that if wolverines were ever native to Michigan, they were extirpated about 200 years ago.

At the time of the wolverine's sighting, DNR Director Rebecca Humphries signed an emergency order protecting the animal from harassment or harm. The animal, which has been seen, photographed and videoed by numerous people since it was discovered, was thought to be alive and well until it was discovered dead Saturday.

The DNR is committed to the conservation, protection, management and accessible use of the state's environment, natural resources and related economic interests for current and future generations.

Meanwhile, the body had been sent to the Michigan DNR's Wildlife Disease Laboratory in Lansing for necropsy, the animal version of an autopsy. Her initial reported weight of 30 pounds—astounding for an adult female wolverine—was attributed to her waterlogged condition and was later adjusted to 25 pounds, and her skinned weight was placed at 15.85 pounds in the necropsy report. The pelt was sent to a taxidermist for preparation as a wildlife display mount.

Much to his surprise and delight, Jeff's request to be present at the necropsy was approved by the state wildlife lab, courtesy of Arnie Karr. As he had throughout her lifetime, Jeff faithfully and painstakingly made a final videotape of the Thumb wolverine.

Tooth analysis indicated that she was about nine years old at the time of death. The solitary state of her existence was also confirmed: she had never borne young. No identification chip or other obvious evidence of former captivity was found.

At the time of death, she appeared to have aspirated stomach contents into her lungs as the terminal event. Ironically, the actual cause of death was noted as cardiomyopathy—the same degenerative heart condition that afflicted Jeff Ford.

Now, in death, the DNA samples they had worked so tirelessly and fruitlessly to acquire were provided in easy abundance. Steve Noble was allowed to collect hair samples directly from the body, which Jeff sent to Audrey Magoun in Alaska, to be forwarded to another lab for the long-awaited final analysis.

Audrey was willing to help on this in any way she could—she wanted the issue laid to rest as much as anyone. Ever since Jeff's 2008 article had put her in an uncomfortable position between the two apparently contradictory lab reports, all her efforts to explain the subtleties of genetics sampling had failed to resolve the issue. In fact, the situation only seemed to worsen as other media outlets ran with the notion that the Thumb wolverine might be a "local gal," as Jeff had described her.

"The thing is the people who want to get these articles out, it's not a big deal to them but it could make a difference to careers when you're quoted wrong," said Magoun. "Most of us realize this happens, the way laypeople often take little bits of science out there and it kind of grows into something else. Before you know it, it doesn't sound like anything you said or could even recognize."

For more than a year, both Magoun and The Wolverine Foundation had been doing their best at damage control, trying to put an end to the public DNA debate by speaking out in qualified support of the original hypothesis that the Thumb wolverine was most likely of Alaskan origin and therefore had probably been a captive at some point that had arrived in Michigan by artificial means. It wasn't that either geneticist had been wrong, she said; the two labs had simply been working with two different databases of comparative samples.

The first tests had determined that the Thumb wolverine's DNA possessed the marker Haplotype C, which was known to exist in roughly one-third of the Alaskan wolverine population and had never been found in wolverines outside that area. The scientists had therefore concluded it was likely that she or her parents had come from Alaska, although they had taken care to point out that there hadn't been much testing for Haplotype C outside the Yukon so there was no way to be sure that it truly was unique to wolverines of Alaskan heredity.

Conversely, the second tests had used an extensive data base of 226 wolverine samples found throughout the populations in Ontario, Manitoba, and up to the Northwest Territories but lacked any Alaskan samples for comparison. Within those samples, the Thumb wolverine's DNA more closely matched that of animals in Ontario and Manitoba than it did those from farther west—so the scientists concluded it was indeed possible she'd come from Ontario stock.

The scientists involved knew neither lab could definitively say which population was the perfect match for the Thumb wolverine's DNA. Without all the parameters involved in one comprehensive analysis, they'd be jumping to unsupported conclusions. They understood the subtle intricacies of those distinctions and why their tests produced what appeared to be conflicting results.

But, like the children's game of "telephone," the fine points of scientific analysis often get lost in translation as the complex information makes its way through the popular media. And in the case of reporters in the "Wolverine State" of Michigan, wishful thinking undoubtedly helped fuel those distortions. Simply put, the idea of a lone wild wolverine making its way across hundreds of miles and a bridge of ice was far more appealing than the theory that one had escaped captivity or been dumped. And the third possibility—that an actual native wolverine population in Michigan had somehow escaped detection until 2004—was for some people the most appealing premise of all.

A third analysis was now under way using the postmortem DNA samples, but by this time the whole affair, quite frankly, had left a bad taste in everyone's mouth.

"We feel the outcomes of the first two analyses were different because only one of the labs involved had Alaskan samples in its database. At this point, TWF opted to accept the analysis of the lab that included the Alaskan samples," said Judy Long of The Wolverine Foundation. "The unfortunate thing is that to say anything seems to somehow put one or the other in a bad light."

Neither the Longs nor Audrey Magoun blamed Jeff for any misconceptions in the press, but by the time all of it had played out, there was simply too much conflicting information circulating to straighten it out.

On July 27, 2010, Joanna Zigouris, a doctoral candidate in the Environmental and Life Sciences Graduate Program at Ontario's Trent University, responded to Arnie Karr's query concerning the status of her expanded analysis of the Thumb wolverine's genetics, hoping some definitive resolution might be provided before the DNR display was ready for the public. Zigouris explained that she would be sequencing additional wolverine samples from across Canada in hopes of determining whether the Haplotype C marker observed in the Yukon was present in a wider range than was currently believed. Until the work was complete, no additional information could be provided.

But even then, would it be a final, definitive answer? No one seemed sure.

On March 30, 2011, the DNR issued this press release.

Michigan's only known wild wolverine is now on display at the visitor center at Bay City State Recreation Area.

The animal was found dead by hikers last winter at Sanilac County's Minden State Game Area, where it had lived for much of the previous six years. The wolverine was first discovered by coyote hunters who treed it while running hounds near Bad Axe on Feb. 24, 2004.

It was the first wolverine ever verified as living in the wild in Michigan. Michigan is known as the Wolverine State because it was a center for trade in the early trapping industry and wolverine pelts from the north and west of Michigan came through the state. Biologists say that if wolverines were native to Michigan, they were extirpated about 200 years ago.

It is uncertain how the wolverine arrived in Michigan, though DNA evidence indicates it is related to animals native to Alaska.

The wolverine was mounted by Bay Port taxidermist Sandy Brown; the mount recently won an award from the state's taxidermy association.

Park interpreter Valerie Blashcka said the display has become quite an attraction.

"It's bringing a lot of visitors who have never been here before," she said. "It's really exciting."

The visitor center, located at 3582 State Park Dr., is open Tuesday through Friday from 10 a.m. to 5 p.m. and Saturday and Sunday noon to 5 p.m.

The Michigan Department of Natural Resources is committed to the conservation, protection, management, use and enjoyment of the state's natural and cultural resources for current and future generations. For more information, go to www.michigan.gov/dnr.

Today, hundreds of visitors have had the opportunity to see Michigan's lone wolverine. Opinions will continue to be expressed and new debates may arise as further findings are presented. But the riddle of her origins may never be fully answered.

Despite the opinions of his colleagues to the contrary, Jeff Ford still maintains his belief that the wolverine made her way to Michigan on her own power from the wilds of northern Ontario.

"I understand that Alaskan samples were not included in the last two analyses, but what I learned with microsatellite markers is the further the sampling is from its original source, the more the subspecies becomes genetically different when considering the microsatellite DNA influence, and Alaska is even further from the 79 percent Ontario match than the Northwest Territories or any other Northwestern province of Canada," said Ford. "Testing using microsatellites in whitetail deer, for example, has shown that although the whitetail deer is all the same species, the DNA geographic branches that are established from the main origin become increasingly varied as the distance from the main 'tree trunk' of microsatellite DNA increases.

"That is why they have found that a deer in northern Saskatchewan, although the same species, has a much different DNA profile than when comparing the DNA profile from a Michigan versus Wisconsin deer. The further the distance, the more distinctly different the DNA microsatellite profile."

In the end, it doesn't really matter, said Judy Long of The Wolverine Foundation. Wherever she came from, however she arrived, or how well she survived, Michigan's lone wolverine has played no small part in saving future generations of her kind.

"The first petition to list the wolverine as an endangered species was in 1994. The final ruling by the US Fish and Wildlife Service came on December 14, 2010, stating the listing was warranted but precluded due to higher priority actions to amend the list," said Long. "Even if the Michigan wolverine's story means nothing more than it was in the limelight receiving a lot of press and causing people to talk about the controversy, it elevates interest in wolverines as a whole. Interest means more funding, and more funding means more opportunities to study this species and make a case for protecting it."

And that's more than enough for any one wolverine—or man—to accomplish in a lifetime. Audrey Magoun forwarded these final words of farewell to Jeff from a friend who maintains the wolverines Audrey raised during her Alaskan research project.

I am so sorry for Jeff. Everyone dreaded this day would come. I know how hard it is to lose a wolverine friend. I hope Jeff can find some consolation in knowing he made her life easier and more interesting for part of her life. She also made a great contribution to people's knowledge and dispelling some of the derogatory myths of wolverines. She was lucky she met Jeff and he has done so much for her and other wolverines by his actions and interest in her. They are both to be commended.

RIP little wolverine.

Notes

✦ ✦ ✦

MARCH 13, 2010

1. V. A. Banci, "Wolverine," in *The Scientific Basis for Conserving Forest Carnivores, American Marten, Fisher, Lynx, and Wolverine in the Western United States,* edited by L. F. Ruggiero, K. B. Aubry, S. W. Buskirk, L. J. Lyon, and W. J. Zielinski, 99–127, USDA Forest Service Rocky Mountain Forest and Range Experiment Station, General Technical Reports, no. RM-254 (Fort Collins, CO: U.S. Forest Service, 1994).

2. Caroline Diem, "Traveling Through Time," *Newberry News* 126, no. 33 (December 28, 2011): 8.

3. Willis F. Dunbar and George S. May, *Michigan: A History of the Wolverine State* (Grand Rapids: William B. Eerdmans, 1995), 215–16.

4. University of Michigan website, "The Origin of the Nickname 'Wolverines,'" http://umgoblue.com/Old/HTML/History/Wolverines.htm.

DECEMBER 18, 1970

1. Ursus International, "Grizzly Bears," retrieved January 30, 2011, from http://www.ursusinternational.org/-current/en/factsgriz.html.

2. US Fish and Wildlife Service, "Living with Grizzlies," retrieved January 30, 2011, from http://www.fws.gov/mountain-prairie/species/mammals/grizzly/grizz _foods.pdf.

FEBRUARY 24, 2004, LATER THAT DAY

1. Michigan Department of Natural Resources, "Gray Wolf (*Canis lupus*)," retrieved February 25, 2011, from http://www.michigan.gov/dnr/0,1607,7-153-10370_12145_12205-32569—,00.html#Michigan%20History.

MARCH 14, 2005

1. US Fish and Wildlife Service press release, December 2010, "Wolverine to Be Designated a Candidate for Endangered Species Protection," retrieved February 26, 2011, from http://us.vocuspr.com/Newsroom/Query.aspx?SiteName= fws&Entity=PRAsset&SF_PRAsset_PRAssetID_EQ=112371&XSL=PressRe lease& Cache=True.

2. Rosemary O'Leary, "Trash Talk: The Supreme Court and the Interstate Transportation of Waste," *Public Administration Review* 57, no. 4 (July–August 1997): 281–84, retrieved January 26, 2011, from http://www.jstor.org/stable/977308.

3. Michigan Department of Environmental Quality, Waste and Hazardous Materials Division, Storage Tank and Solid Waste Section, "Report of Solid Waste Landfilled in Michigan October 1, 2003–September 30, 2004," January 22, 2005, retrieved February 26, 2011, from http://www.michigan.gov/documents/deq/deq-whmd-swp-FY2004-SW-Land‹lled-Rpt_247495_7.pdf.

4. Office of Senator Carl Levin, press release, November 15, 2004, "Number of Canadian Trash Trucks Entering Michigan Doubles in One Year: Stabenow, Levin, Dingell Ask Homeland Security Secretary Tom Ridge for Immediate Action," retrieved January 26, 2011, from http://levin.senate.gov/newsroom/release.cfm?id=227978.

5. "Michigan Is Cougar Country," Michigan Wildlife Conservancy presentation to the Michigan Outdoor Writers Association, February 7, 2004, Lewiston, MI.

6. Michigan Department of Natural Resources, "Michigan Cougar History," retrieved January 28, 2011, from http://www.michigan.gov/dnr/0,1607,7-153-10370_12145_43573-153226—,00.html.

7. Michigan Wildlife Conservancy, "About Us Biographies," retrieved January 28, 2011, from http://www.miwildlife.org/bio-dennis.asp.

8. Ibid.

9. Tom Carney, "Michigan's Cougar Saga: A Lesson in Natural Resources Reporting," presentation to the Michigan Outdoor Writers Association, February 7, 2004, Lewiston, MI.

10. Victor Skinner, "Photo Shows Cougar Presence in Michigan," *Grand Rapids Press*, November 15, 2009, retrieved January 29, 2011, from http://www.mlive.com/outdoors/index.ssf/2009/11/photo_shows_cougar_presence_in.html.

11. Randy Conat, "DNR Says No Cougars in Lower Peninsula," *ABC News-12* (Flint, Michigan), January 7, 2009, retrieved January 30, 2011, from http://abclocal.go.com/wjrt/story?section=news/local&id=6591611.

LATE SPRING, 2005

1. Michigan Department of Natural Resources, "Bald Eagle (*Haliaeetus leucocephalus*) Life History," retrieved February 6, 2011, from http://www.michigan.gov/dnr/ 0,1607,7-153-10370_12145_12202-32581—,00.html.

2. The Wolverine Foundation, TWF home page, retrieved May 2, 2011, from http://www.wolverinefoundation.org.

3. Michigan Department of Natural Resources, "Michigan Black Bear Facts," retrieved February 20, 2011, from http://www.michigan.gov/dnr/0,1607,7-153-10369-105034—,00.html.

4. "Ontario's Crown Land Use Policy Atlas," retrieved February 20, 2011, from http://crownlanduseatlas.mnr.gov.on.ca/supportingdocs/alus/contents .htm.

5. Committee on the Status of Endangered Wildlife in Canada, "COSEWIC Assessment and Update Status Report on the Wolverine, 2003," retrieved February 20, 2011, from http://dsp-psd.pwgsc.gc.ca/Collection/CW69-14-329-2003E .pdf.

6. MODIS Rapid Response Team, NASA, "Ice Covers the Great Lakes," retrieved October 29, 2011, from http://earthobservatory.nasa.gov/IOTD/view .php?id=3280.

7. National Ice Center, "Great Lakes Ice Analysis Products," retrieved October 29, 2011, from http://www.natice.noaa.gov/pub/special/great_lakes/2004/charts/composite_east/el040219color.jpg.

8. The Wolverine Foundation, "Wolverine Project: Ecology of Captive-Born Kits Raised in Natural Habitat," retrieved March 26, 2011, from http://wolverine foundation.org/729-2/.

FEBRUARY, 2006

1. Robert M. Inman et al., "Wolverine Makes Extensive Movements in the Greater Yellowstone Ecosystem," *Northwest Science* 78, no. 3 (2004): 261–66.

2. US Fish and Wildlife Service, "Wolverine Fact Sheet," Retrieved March 13, 2011, from http://www.fws.gov/mountain-prairie/species/mammals/wolverine/wolverine-122010.pdf.

3. Inman et al., "Wolverine Makes Extensive Movements in the Greater Yellowstone Ecosystem."

4. Ontario Ministry of Natural Resources, "Ontario's Forests: The Boreal Forest," retrieved March 13, 2011, from http://www.mnr.gov.on.ca/en/Business/Forests/2ColumnSubPage/240961.html.

5. Michigan Department of Natural Resources, "Remembering Michigan's Historic Moose Lift," retrieved March 13, 2011, from http://www.michigan.gov/dnr/0,1607,7-153-10366_46403_46404-110006—,00.html.

6. Michigan Department of Natural Resources, "Restoration of the Wild Turkey Is a Wildlife Success Story," retrieved March 13, 2011, from http://www.michigan.gov/dnr/0,1607,7-153-10366_46403_46404-213508—,00.html.

7. Minnesota Climatology Working Group,"Lake Superior Freeze Over? March 6, 2003," retrieved March 13, 2011, from http://climate.umn.edu/doc/journal/superior030603.htm.

8. NASA Visible Earth, "Frozen Lake Superior, March 12, 2003," retrieved March 13, 2011, from http://visibleearth.nasa.gov/view_rec.php?id=5185.

APRIL, 2006

1. Bryn Mickle, "3 Closer to Trial in Animal Thefts," *Flint Journal*, November 20, 2007, retrieved March 26, 2011, YPERLINK"http://blog.mlive.com/flintjournal/newsnow/2007/11/3_closer_to_trial_in_animal_th.html"http://blog.mlive.com/flintjournal/newsnow/2007/11/3_closer_to_trial_in_animal_th.html.

2. Roneisha D. Mullen, "Adam Lock Sent to Jail," *Flint Journal*, September 23, 2008, retrieved March 26, 2011, http://www.mlive.com/news/flint/index.ssf/2008/09/adam_lock_sent_to_jail_for_11.html.

2008

1. Thomas P. Sullivan, "Influence of Wolverine (*Gulo gulo*) Odor on Feeding Behavior of Snowshoe Hares (*Lepus americanus*)," *Journal of Mammalogy* 67, no. 2 (May 1986): 385–88, retrieved from http://www.jstor.org/pss/1380892.

2. "Student Captures Image of Rare Wolverine in California," *Science Daily*, March 9, 2008, retrieved April 10, 2011, from http://www.sciencedaily.com/releases/2008/03/080306144240.htm.

Index

✦ ✦ ✦

Text design by Jillian Downey
Typesetting by Delmastype, Ann Arbor, Michigan
Text font: Janson
Display font: TheSans

Although designed by the Hungarian Nicholas Kis in about 1690, the model for Janson Text was mistakenly attributed to the Dutch printer Anton Janson. Kis' original matrices were found in Germany and acquired by the Stempel foundry in 1919. This version of Janson comes from the Stempel foundry and was designed from the original type; it was issued by Linotype in digital form in 1985.

—courtesy www.adobe.com

Lucas de Groot, of the LucasFonts foundry, created TheSans in 1994, as part of his Thesis family of fonts. The letters have a diagonal stress and a forward flow that facilitates reading.

—courtesy www.lucasfonts.com